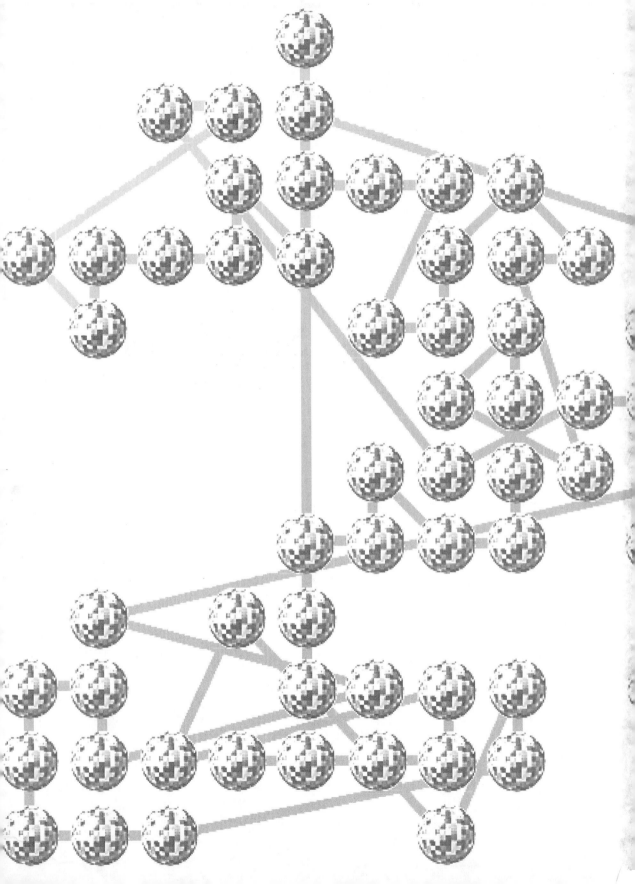

Computer Simulations with MATHEMATICA®

Explorations in Complex Physical and Biological Systems

RICHARD J. GAYLORD

PAUL R. WELLIN

SPRINGER-VERLAG · TELOS · THE ELECTRONIC LIBRARY OF SCIENCE

Richard J. Gaylord
Dept. of Materials Science
    and Engineering
University of Illinois,
Champaign-Urbana
Urbana, IL 61801 USA

Paul R. Wellin
Dept. of Mathematics
California State University, Sonoma
Rohnert Park, CA 94928 USA

Published by TELOS®, The Electronic Library of Science, Santa Clara, CA
Publisher: Allan M. Wylde
Publishing Assistant: Kate McNally Young
Product Manager: Carol Wilson
Production and Manufacturing Manager: Jan V. Benes
Electronic Production Advisor: Joe Kaiping
Cover and Page Designer: Peter Tucker

12-4-95 - 1864610

Gaylord, Richard J.
        Computer simulations with Mathematica: explorations in the physical, biological, and
    social sciences / Richard J. Gaylord, Paul R. Wellin.
            p. cm.
        Includes bibliographical references and index.
        ISBN 0-387-94274-2
        1. Digital computer simulation.   2. Mathematica (Computer file).   I. Wellin, Paul R.   II. Title.
    QA76.9.C65G39   1994 *1995*
    003' .353—dc20   94-17623
                                                                         CIP

The use of general descriptive names, trademarks, etc., in this publication, even if the former are not
especially identified, is not to be taken as a sign that such names, as understood by the Trade Marks and
Merchandise Marks Act, may accordingly be used by anyone. Where those designations appear in this
book and Springer-Verlag was aware of a trademark claim, the designations follow the capitalization style
used by the manufacturer. Apple and Macintosh are registered trademarks of Apple Computer Inc.;
Mathematica, Mathlink, and MathSource are registered trademarks of Wolfram Research, Inc.; Microsoft
Windows and MS-DOS are registered trademarks of Microsoft Corporation; NeXT is a registered trade-
mark of NeXT Computer Inc.; PostScript is a registered trademark of Adobe Systems, Inc.; Sun is a trade-
mark of Sun Microsystems, Inc.; TEX is a registered trademark of the American Mathematical Society;
UNIX is a registered trademark of AT&T. All other product names are trademarks of their producers.
Mathematica is not associated with Mathematica, Inc.; Mathematica Policy Research Inc.; MathTech, Inc.

Photocomposed pages prepared from the author's TEX files.

9 8 7 6 5 4 3

ISBN: 0-387-94274-2 Springer-Verlag New York Berlin Heidelberg
ISBN: 3-540-94274-2 Springer-Verlag Berlin Heidelberg New York

TELOS, The Electronic Library of Science, is an imprint of Springer-Verlag New York with publishing facilities in Santa Clara, California. Its publishing program encompasses the natural and physical sciences, computer science, economics, mathematics, and engineering. All TELOS publications have a computational orientation to them, as TELOS' primary publishing strategy is to wed the traditional print medium with the emerging new electronic media in order to provide the reader with a truly interactive multimedia information environment. To achieve this, every TELOS publication delivered on paper has an associated electronic component. This can take the form of book/diskette combinations, book/CD-ROM packages, books delivered via networks, electronic journals, newsletters, plus a multitude of other exciting possibilities. Since TELOS is not committed to any one technology, any delivery medium can be considered.

The range of TELOS publications extends from research level reference works through textbook materials for the higher education audience, practical handbooks for working professionals, as well as more broadly accessible science, computer science, and high technology trade publications. Many TELOS publications are interdisciplinary in nature, and most are targeted for the individual buyer, which dictates that TELOS publications be priced accordingly.

Of the numerous definitions of the Greek word "telos," the one most representative of our publishing philosophy is "to turn," or "turning point." We perceive the establishment of the TELOS publishing program to be a significant step towards attaining a new plateau of high quality information packaging and dissemination in the interactive learning environment of the future. TELOS welcomes you to join us in the exploration and development of this frontier as a reader and user, an author, editor, consultant, strategic partner, or in whatever other capacity might be appropriate.

TELOS, The Electronic Library of Science
Springer-Verlag Publishers
3600 Pruneridge Avenue, Suite 200
Santa Clara, CA 95051

TELOS Diskettes

Unless otherwise designated, computer diskettes packaged with TELOS publications are 3.5" high-density DOS-formatted diskettes. They may be read by any IBM-compatible computer running DOS or Windows. They may also be read by computers running NEXTSTEP, by most UNIX machines, and by Macintosh computers using a file exchange utility.

In those cases where the diskettes require the availability of specific software programs in order to run them, or to take full advantage of their capabilities, then the specific requirements regarding these software packages will be indicated.

TELOS CD-ROM Discs

It is also clearly indicated to buyers of TELOS publications containing CD-ROM discs, or in cases where the publication is a standalone CD-ROM product, the exact platform, or platforms, on which the disc is designed to run. For example, Macintosh only; MPC only; Macintosh and Windows (cross-platform), etc.

TELOSpub.com (Online)

New product information, product updates, TELOS news and FTPing instructions can be accessed by sending a one-line message: **send info** to: **info@TELOSpub.com**. The TELOS anonymous FTP site contains catalog product descriptions, testimonials and reviews regarding TELOS publications, data-files contained on diskettes accompanying the various TELOS titles, order forms and price lists.

To Carole

—RJG

To my parents,
and their parents before them …

—PRW

Preface

Doing Science with Algorithms

The goal of scientific investigation is to understand how nature works. There are two ways to do this: by experimental observation and by theoretical modeling. Modeling has traditionally consisted of thinking up and writing down equations and then solving those equations, either by hand (*e.g.*, theoretical physics) or by computer (*e.g.*, computational physics). While the use of equations has been very useful in opening up the natural world to our understanding, as well as to our control, it has its limitations. The equational approach breaks down when the equations can't be solved (analytically or numerically) because of technical difficulties or when the system being studied cannot be represented in terms of equations.

A new approach to the modeling of natural phenomena does not use equations at all. In *algorithmic physics*, as we call this approach, equations are replaced by algorithms or computer programs. These programs, known as computer simulations, directly model the phenomenon under investigation. Simulations are especially useful for studying complex systems.

In this book, we show how to create computer simulation programs and how to visually and numerically analyze the results obtained by running these programs. In order to show the breadth of the method in the study of natural phenomena, we look at various physical, chemical, biological, and social systems.

Doing Simulations with *Mathematica*

The computer simulation programs in this book are written in the *Mathematica* programming language. This language is a natural choice because, as its principal creator, Stephen Wolfram, has stated, *Mathematica* is designed for use as a computational tool for scientific exploration. In particular, *Mathematica* is used here for the following reasons:

- *Mathematica* is a very powerful high-level programming language. That is to say, computations are written in, and performed by, a *Mathematica* program in almost exactly the same way as the user would express and perform them.

- The *Mathematica* language is very simple, making it easy to understand and to master, even for the "non-hacker" who views programming as a tool rather than an end in itself.
- *Mathematica* has extensive, easy-to-use graphics capabilities which make it possible to use the tools of scientific visualization to analyze computer simulation results, and also to debug programs.
- *Mathematica* has extensive numeric capabilities which make it possible to numerically analyze computer simulation results.
- All of the above items are wrapped in an integrated computing environment with an extremely well designed user interface known as a *notebook*.
- *Mathematica* can be run using a local front end on one computer and a remote kernel on another platform. It can also be linked to other programming languages and software.

On a personal note, this book is intended for the enjoyment, as well as the edification, of the reader. Our purpose is to convey the pleasures of both computer simulation and *Mathematica* programming. If, after going through this book, you not only have a sense of the enormous potential of computer simulation as a tool for exploring the world around you but, moreover, feel that you want to carry out these explorations on your own, that will be especially gratifying to us.

ACKNOWLEDGMENTS

We wish to thank a number of people who have contributed to make this book possible. Our publisher, Allan Wylde must be thanked for his continuing encouragement and publication of materials with significant electronic components. Jan Benes helped steer the project through the production process, adeptly navigating a very close approximation to a shortest path. Joe Kaiping, as always, provided exceptional TEX macros and support.

Richard J. Gaylord would like to acknowledge a number of people who have contributed in various ways to this book.

Bill Tyndall, my friend since way back when (the seventh grade to be precise) was with me on my maiden voyages into writing computer simulations with *Mathematica* and he helped me to get my programming "sea legs."

Kazume Nishidate, a colleague and friend of mine whom I first met in cyberspace, has been my traveling companion into the world of computer simulations. We work together via the Internet (the fact that the last portion of his day in Japan coincides with the first portion of my day in Illinois enables us to work together interactively) and some of the programs in this book were developed in collaboration with Kazume.

The students in my Materials Science and Engineering 382 computer simulation course in Fall 1993 challenged me with alternative programming approaches to some of my cellular automata systems. Whenever their approaches turned out to be more elegant or more efficient than mine, I adopted them with some modifications and the results have found their way into this book.

My cardiologist, Leslie Fleischer, provided excellent medical care that made it possible for me to do this book and I hope that we'll continue our wide-ranging discussions of medicine, politics, food, computers, and movies for years to come.

The hospitality extended to me by Professor Sir Sam Edwards, the Cavendish Laboratory, and my college, Clare Hall, at the University of Cambridge and the use of their facilities while this book was being completed was much appreciated.

Paul Wellin would like to thank Tom Wickham-Jones for helping him to understand much of what underlies the graphics routines in *Mathematica*.

Paul Wellin also thanks Sheri, Sam, and Oona for their support and encouragement throughout this project. An author often comes to the conclusion that writing a book is best accomplished by going into seclusion for periods of time. Thanks to my family for providing a system of checks and balances.

Richard J. Gaylord
Paul R. Wellin
October 1994

How to Use This Book

This book is intended to be used as a "how to" manual, providing an entry for you, the reader, into the world of computer simulations. Accordingly, the topics covered are rather wide-ranging so that you may find several subjects of interest. The relatively short headings in the Table of Contents cannot fully indicate the relevance of each chapter to various fields, so it is a good idea to start by reading the brief introductory section of each chapter which points out the natural phenomena for which the ensuing computer simulation model is relevant.

The appendix on *Mathematica* programming should be perused before starting in on any of the chapters. This is a self-contained explanation of how the *Mathematica* programming language works, and should make you feel more comfortable with the language. The chapters of this book can then be read systematically or in a "pick-and-choose" fashion, as suits your interests. Each chapter is organized in a similar manner:

- The brief description at the start of a chapter is intended simply to indicate the usefulness of the simulation model that follows. To learn more about the scientific and historical background of the model, you should read the articles referenced at the end of the chapter. Those articles marked with an asterisk offer an overview of the subject for the nonspecialist. The other articles are more technical and contain references to the research literature.

- Following the introduction, the algorithm for the computer simulation model is stated and then the main task of the chapter is undertaken: a demonstration of how to write a program to implement the algorithm. The program is developed in a step-by-step manner, with detailed explanations of the individual code fragments that make up the program. You can, and should, enter these fragments with simple argument values to see them work.

- After the simulation program is created, additional programs are presented which can be used to analyze numerically and graphically the results obtained from running the simulation program. Some typical results from running these programs are usually shown for rather small-sized systems. In many cases a simulation of a large-sized system can be found on the CD-ROM that accompanies this book.

- Each chapter ends with a projects section containing various programs related to the subject of the chapter and suggestions for other computations. This material is an integral part of the chapter and can provide a launching point for you to begin your own explorations of the subject.

Finally, it should be noted that, in order to increase the utility of the *Mathematica* programming language for writing computer simulation (as well as other) programs, the functional style of programming is emphasized in this book. In keeping with this approach (1) looping is mostly avoided, (2) conditional branching is minimized, (3) data structures, such as lists, are manipulated in their entirety rather than in a piecemeal fashion, (4) built-in *Mathematica* functions are employed whenever possible, and (5) *anonymous* functions, *higher-order* functions and *nested* function calls are used extensively. The rule-based programming style is also emphasized through the use of look-up rule tables employing pattern-matching, rather than computation, whenever practical. These programming techniques (and the detailed explanations of the programs) should make the book useful not only to those interested in computer simulations, but to anyone who wants to learn how to write elegant and efficient *Mathematica* programs for whatever purpose.

MATHLINK AND REMOTE COMPUTING

Computing has come a long way since the days of punchcards and weekend batch jobs. The computational speed of workstations, desktop and even laptop computers of today dwarfs the computers of only a decade ago. Yet with every increase in raw computational power comes our desire to solve still more complicated problems that require ever more machine resources.

The interpreted nature of the *Mathematica* programming language allows for a natural development of computer simulation programs, avoiding the "compile" and "make" contortions required by other languages. This is an ideal environment for prototyping simulation programs and is also sufficient for running many of the simulations in this book. Occasionally, however, you may find that your computer lacks the speed or memory necessary to run a simulation of a sufficiently large system over a long enough period of time to see certain patterns or behavior emerge. Fortunately, you have two solutions to this dilemma.

If you have access to a remote computer on which *Mathematica* is installed, you can connect your machine to the remote computer, perform the computations on that remote machine, and then use your results locally—graphing them or analyzing them numerically, for example. You can either connect to a machine on your network, or send your code to a remote machine off-site and bring back the results when done. Both of these processes are described in Appendix D.

Even if you cannot connect your computer to a remote machine, you can greatly increase the running times of many of your *Mathematica* programs by casting some of the computations in a compiled language such as C or Fortran. In this case, you write a C program, say, compile it in your operating system and then "call" that external program from within *Mathematica*. The communication between *Mathematica* and your external program is handled seamlessly by the *MathLink* protocol which ships with each copy of *Mathematica*. Appendix C on *MathLink* describes how this process can be done on a variety of machines. In addition, Appendix E includes a listing of all of the C source code described in this book. All source code (and executables for Windows and Macintosh platforms) are included in electronic form on the CD-ROM for you to use and modify for your own purposes.

THE ELECTRONIC COMPONENT

This book includes a CD-ROM that contains all the material in the book plus additional graphics images, animations, and movies of the simulations. Each chapter is included as a *Mathematica* notebook that allows you to "step through" the algorithms in an interactive manner. The simulations on the CD-ROM are run for many more parameter values and generally are run on larger grids which often show clearer pattern formation and much higher resolution. Numerous animations and movies are included on disk to help see the dynamic process involved in many of the simulations discussed in the book.

The CD-ROM disk will automatically mount on Macintosh, DOS, Windows, and NEXTSTEP operating systems by inserting the CD-ROM into your CD-ROM drive. Some Unix users will have to manually mount the CD-ROM. Instructions for mounting the disk on DEC, HP, RISC, SGI, and SUN machines are included with the CD-ROM.

Utilities are included for viewing the graphics files and the QuickTime movies. In addition, *Mathematica* notebooks can be viewed and printed (even if you do not own a copy of *Mathematica*) with the versions of MathReader included on the CD-ROM.

Contents

Preface   *ix*
How to Use This Book   *xiii*

PART I *Probabilistic Systems*

1. **The Random Walk**   **3**
One- and two-dimensional walks, calculating the mean shape of
a walk, determining the critical exponent, visualizing the walk

2. **The Self-Avoiding Walk**   **17**
Slithering snake motion, rearrangement by pivoting

3. **Accretion**   **31**
Diffusion-limited aggregation, determining the fractal dimension, ballistic deposition, visualizing DLAs

4. **Spreading Phenomena**   **45**
Single cluster growth, invasion percolation

5. **Percolation Clustering**   **57**
Random site percolation, cluster labeling with the Hoshen-Kopelman
algorithm, continuum percolation

6. **The Ising Model**   **75**
Probabilistic Ising model, the Metropolis method, magnetization
behavior

7. *Darwinian Evolution*　　　　　　　　　　　　　　*83*
Co-evolution and punctuated equilibrium

PART II *Cellular Automata*

Cellular Automata Preliminaries　　　　　　　　　*91*
Defining a cellular automaton, lattices, neighborhoods, boundary conditions

8. *The Game of Life*　　　　　　　　　　　　　　*97*
Game of Life, diffusion, boiling and weathering cellular automata

9. *Avalanches*　　　　　　　　　　　　　　　*105*
Self-organized criticality, sandpiles

10. *The Q2R Ising Model*　　　　　　　　　　　*111*
Ising cellular automaton

11. *Excitable Media*　　　　　　　　　　　　*117*
Self-propagating patterns, neuron action, cyclic space cellular automata, the hodgepodge machine for oscillatory chemical reactions

12. *Traffic*　　　　　　　　　　　　　　　*135*
One-lane traffic with car stopping, two-lane one-way road with car passing, accidents and road work, the fundamental diagram

13. *Forest Fires*　　　　　　　　　　　　　*147*
Deforestation, reforestation, forest size distribution

14. *Complexity*　　　　　　　　　　　　　*157*
One-dimensional cellular automata, Wolfram rules, animations

PART III *Appendices*

A. *Mathematica Programming*　　　　　　　　*173*
Expressions, patterns, evaluation, rewrite rules, transformation rules, higher-order functions

B. **Random Numbers**   *207*
Random number generators, tests for randomness, using different probability distributions

C. **Computer Simulations and MathLink**   *221*
Using *MathLink* to call external programs from within *Mathematica*, techniques in *MathLink* programming, by Todd Gayley

D. **Remote Computing with Mathematica**   *255*
Connecting to a remote computer with a local front end, computing across networks

E. **MathLink Program Listing**   *263*
A listing of all *MathLink* programs including Walk2DC, SeedRandomC, PhasesC, SandpileC, OneLaneC, EpidemicC, and LifeGameC

Index   *291*

CD-ROM Contents

Mathematica Notebooks
This directory contains *Mathematica* notebooks of the text and code from each chapter and appendix, special notebooks for running the graphics from the book, and *MathLink* notebooks that show you how to install and run the *MathLink* code.

Graphics Images
Single frame graphics images of the simulations are included in a variety of formats. Both black and white and color images are included.

Animations
QuickTime movies and *Mathematica* notebooks containing animations are included.

C Code
The code for each of the *MathLink* functions is included. The code is generic and will run on Macintosh, Windows, and Unix platforms.

Binaries
Binary files are included for the Macintosh and DOS/Windows platforms.

Utilities
Utilities for each of the platforms are included to allow you to view the graphics, animations and movies, and *Mathematica* notebooks.

Probabilistic Systems

CHAPTER 1

The Random Walk

INTRODUCTION

Envision the following process: A person takes a step in a randomly chosen direction, then takes another step in a randomly chosen direction, and then takes another step in a randomly chosen direction, and so on until a total of n steps have been taken. This process is known as the random walk model and it has been extremely useful to scientists in many fields who study stochastic (probabilistic) processes: physicists modeling the transport of molecules, biologists modeling the locomotion of organisms, and economists modeling the behavior over time of financial markets. In fact, the very first applications of the random walk model were made in these areas at the beginning of this century. The widespread use of the random walk model, and its simple description and implementation, make it a good jumping-off point for our explorations in the world of computer simulation.

1.1 RANDOM WALK PROGRAMS

THE ONE-DIMENSIONAL RANDOM WALK

The simplest random walk model consists of n steps of equal length, back-and-forth along a horizontal line. A step (or step increment) in the positive x-direction corresponds to a value of 1 and a step in the negative x-direction corresponds to a value of -1. A list of the successive step increments of an n-step random walk in one dimension is therefore a list of n randomly selected 1s and -1s. This list can be generated in many ways. We will use the following construction:

```
Table[(-1)^Random[Integer], {n}]
```

Using this `Table`, we can write a program, called `StepIncrements`, which generates a sequence of n step increments.

```
In[1]:= StepIncrements[n_] := Table[(-1)^Random[Integer], {n}]
```

A typical run of the `StepIncrements` program where $n = 10$ is shown below.

```
In[2]:= StepIncrements[10]

Out[2]= {-1, 1, -1, -1, -1, -1, 1, -1, 1, -1}
```

A list generated with `StepIncrements[n]` can be used to generate a list of the $(n + 1)$ locations of a one-dimensional n-step walk which starts at the origin, using the `FoldList` function. For example, using the result just obtained, we can generate a 10-step walk as follows:

```
In[3]:= FoldList[Plus, 0, %]

Out[3]= {0, -1, 0, -1, -2, -3, -4, -3, -4, -3, -4}
```

Using the right hand side of the `StepIncrements` program in the `FoldList` function, we can write the program `Walk1D` to generate a list of the step locations of an n-step random walk, starting at the origin.

```
In[4]:= Walk1D[n_] :=
          FoldList[Plus, 0, Table[(-1)^Random[Integer], {n}]]
```

Here is a typical run of the `Walk1D` program where $n = 10$.

```
In[5]:= Walk1D[10]

Out[5]= {0, 1, 2, 3, 2, 1, 2, 1, 2, 3, 2}
```

Note: A list of the step locations can also be generated, albeit a bit more slowly, without first creating a list of the step increments. This can be accomplished by using a nesting operation with the built-in *Mathematica* function `NestList`.

```
NestList[(# + (-1)^Random[Integer])&, 0, n]
```

In the next section we'll look at the random walk model in higher dimensions.

THE TWO-DIMENSIONAL LATTICE WALK

The random walk model in higher dimensions is a bit more complicated than the random walk in one dimension. In one dimension, each step of the walk is either 0 degrees (a forward step) or 180 degrees (a backward step) with respect to the preceding step. In higher dimensions, a step can take a range of orientations with respect to previous steps.

We'll consider a random walk on a lattice. This kind of walk is appropriately referred to as a *lattice walk*. Specifically, we'll look at a lattice walk on the two-dimensional square lattice. This walk consists of steps of uniform length, randomly taken in the north, east, south, or west direction. The list of the possible increments for a step in this walk are then given by the following list.

```
{{0, 1}, {1, 0}, {0, -1}, {-1, 0}}
```

A list of n step increments can be created from this list as follows:

```
{{0,1}, {1,0}, {0,-1}, {-1,0}}[[Table[Random[Integer,
                        {1,4}], {n}] ]]
```

By analogy to the one-dimensional walk computation, here is a program, `Walk2D`, that can be used to generate a list of the step locations of an n-step lattice walk starting at the origin $\{0, 0\}$.

```
In[6]:= Walk2D[n_] :=
            FoldList[Plus, {0,0},
                    {{0,1}, {1,0}, {0,-1}, {-1,0}}[[
                Table[Random[Integer, {1,4}], {n}] ]]
            ]
```

A typical run of the `Walk2D` program is shown below where $n = 10$.

```
In[7]:= Walk2D[10]

Out[7]= {{0, 0}, {1, 0}, {2, 0}, {2, -1}, {1, -1}, {1, -2},
            {1, -3}, {1, -2}, {1, -3}, {1, -2}, {1, -3}}
```

1.2 ■ NUMERICAL ANALYSIS OF THE TWO-DIMENSIONAL LATTICE WALK

In studying a process that is random in nature, we are often interested in its *mean*, or *average*, properties. To obtain the mean value of a quantity, the quantity is computed a number of times, the values that are obtained are summed, and the result is divided by the number of computations. We'll look at two measures of the mean "size" of a two-dimensional lattice walk: the *mean square end-to-end distance* and the *mean square radius of gyration.*

MEAN SQUARE END-TO-END DISTANCE

The square end-to-end distance r^2 of a two-dimensional lattice walk is given by $(x_f - x_i)^2 + (y_f - y_i)^2$ where $\{x_i, y_i\}$ and $\{x_f, y_f\}$ are the initial and final locations of the walk, respectively. Choosing the origin $\{0, 0\}$ as the starting point of the lattice walk simplifies the formula to $(x_f^2 + y_f^2)$.

The square end-to-end distance of a lattice walk starting at $\{0, 0\}$ and ending at $\{x_f, y_f\}$ is given by the following:

```
Apply[Plus, {xf, yf}^2]
```

Here is a program `SquareDistance` that uses the above formula to compute r^2 for the two-dimensional lattice walk. (Note: `Last[walk].Last[walk]` could also be used to compute `SquareDistance[walk]`.)

```
In[1]:= SquareDistance[walk_List] := Apply[Plus, Last[walk]^2]
```

A typical run of the `SquareDistance` program is shown below for a 10-step random walk.

```
In[2]:= SquareDistance[Walk2D[10]]

Out[2]= 36
```

A program for computing the mean square, end-to-end distance $<r^2>$ for m n-step lattice walks can be written very easily, using the `Sum` function.

```
In[3]:= MeanSquareDistance[n_Integer, m_Integer] :=
           Module[{Walk2D},
             Walk2D[s_] := FoldList[Plus, {0,0},
                {{0,1}, {1,0},
                 {0,-1}, {-1,0}}[[Table[Random[Integer,
                                    {1,4}], {s}]]]];
             N[Sum[Apply[Plus, Last[Walk2D[n]]^2], {m}]/m]
           ]
```

A typical run of the `MeanSquareDistance` program is shown below for 25 10-step walks.

In[4]:= **MeanSquareDistance[10, 25]**

Out[4]= 8.08

MEAN SQUARE RADIUS OF GYRATION

The mean square radius of gyration $< R_g^2 >$ of a random walk is the sum of the squares of the distances of the step locations from the center of mass, divided by the number of step locations, where the center of mass is the sum of the step locations divided by the number of step locations.

The computations of the center of mass and the sum of the squares of step distances from the center of mass, are relatively straightforward.

For the list of the $(n + 1)$ locations of an n-step walk, which we'll call locs, the center of mass (cm) can be defined as follows:

cm = N[Apply[Plus, locs]/(n + 1)]

In the list $\{\{(x_0 - x_{cm})^2, (x_1 - x_{cm})^2, \ldots, (x_n - x_{cm})^2\}, \{(y_0 - y_{cm})^2, (y_1 - y_{cm})^2, \ldots, (y_n - y_{cm})^2\}\}$, x_j and y_j are the x- and y-coordinates of the jth step location. In addition, x_{cm} and y_{cm} are the coordinates of the center of mass location. This list is then given by:

(Transpose[locs] - cm)^2

Hence, the sum of the squares of distances of the step locations from the center of mass can be computed as follows:

Apply[Plus, Flatten[(Transpose[locs] - cm)^2]]

The overall program for the mean square-radius of gyration is written using the above expressions and follows the form used for the MeanSquareDistance program.

```
In[5]:= MeanSquareRadiusGyration[m_Integer, n_Integer] :=
           Module[{squareRadiusGyration},
             squareRadiusGyration[s_Integer] :=
               Module[{locs, cm, choices = {{1,0}, {-1,0},
                                             {0,1}, {0,-1}}},
                 locs = FoldList[Plus,{0,0},
                         choices[[Table[Random[Integer,{1,4}],
                                        {s}]]]];
                 cm = N[Apply[Plus, locs]/(s+1)];
                 Apply[Plus,
                     Flatten[(Transpose[locs] - cm)^2]]/(s+1)
               ];
             N[Sum[squareRadiusGyration[n], {m}]/m]]
```

A typical run of the `MeanSquareRadiusGyration` program is shown below for 25 10-step walks.

```
In[6]:= MeanSquareRadiusGyration[10, 25]

Out[6]= 4.53639
```

THE CRITICAL EXPONENT OF A RANDOM WALK

It has been found experimentally (in the laboratory) and theoretically (on paper) that both of the means $<r^2>$ and $<R_g^2>$ have a power law dependence on the number n of steps in the walk. The power v in the relationship $<r^2>= n^v$ is known as the *critical exponent* of the walk. The computation of the critical exponent can be illustrated with a specific example.

We'll first create a data set of the computed values of the ordered pairs $\{<r^2>,$ $n\}$ for values of n from 10 to 90 in increments of 20, with 75 walks ($m = 75$) being used to compute $<r^2>$ in each case.

```
In[7]:= data = Map[({MeanSquareDistance[#, 75], #})&,
           Range[10, 90, 20]]

Out[7]= {{7.54667, 10}, {33.84, 30}, {47.1467, 50},
         {60.2933, 70}, {96.08, 90}}
```

Here, we take the `Log` of the data (equivalent to creating the list of ordered pairs, $\{$`Log[<r^2>]`, `Log[n]`$\}$), and use it in the `Fit` function:

```
In[8]:= calculatedExponent = Fit[N[Log[data]], {x}, x]

Out[8]= 1.00993 x
```

For a power law relationship of the form $< r^2 >= n^v$, taking logs gives Log[<r^2>] = v Log[n], so the coefficient of x in calculatedExponent is the critical exponent, which has a value of approximately 1.

When we used x in the Fit function, we assumed, rather than proved, a power law dependence of $< r^2 >$ on n. To check how good this assumption is, we can look at plots of the computed power law relationship together with the data set.

Here is a plot of the computed power law relationship.

In[9]:= **calculatedGraph = Plot[calculatedExponent, {x, 0, 90}]**

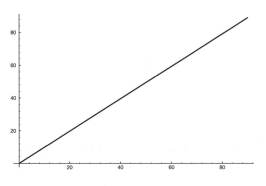

Out[9]= -Graphics-

A log-log plot of the data is created using the LogLogListPlot function defined in the *Mathematica* Graphics package. First we need to load the package and then create the plot dataGraph.

In[10]:= **Needs["Graphics'Graphics'"]**

In[11]:= **dataGraph = LogLogListPlot[data,**
                    PlotStyle -> PointSize[.03]]

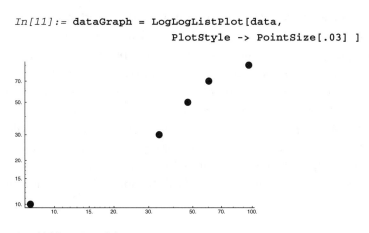

Out[11]= -Graphics-

The plots of the computed power law relationship and the data set can now be shown together using Show.

In[12]:= **Show[dataGraph, calculatedGraph]**

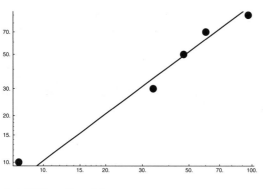

Out[12]= -Graphics-

1.3   VISUALIZING THE TWO-DIMENSIONAL LATTICE WALK

SIMPLE GRAPHICS

We can create a snapshot of the path of the lattice walk using the graphics primitive Line to draw lines between successive points in the walk.

```
In[1]:= ShowWalk2D[coords_, opts___] :=
            Show[Graphics[Line[coords],
                    opts,
                    AspectRatio -> Automatic]]
```

Here we have set the value of the AspectRatio option to Automatic so that steps in the *x*- and *y*-directions have equal lengths. This option can be overwritten by specifying a different value in the list of options given by opts. Note the use of the triple blank in the definition of ShowWalk2D. The pattern opts___ matches any sequence (possibly empty) of rules which are used here to govern the display of the graphic by changing certain options to the Graphics function. It is important that opts appears *before* the option AspectRatio. This will allow you to override this (or any) value because *Mathematica* uses the value for the first option that it encounters if that option is listed more than once. If opts had come at the end of this function, you would not be able to change the value for AspectRatio.

Here is a simple example of a 20-step, two-dimensional lattice walk.

In[2]:= **ShowWalk2D[Walk2D[20]]**

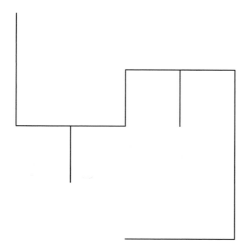

Out[2]= -Graphics-

The next example shows a random walk of length 1000 and uses some additional options to change the color of the line segments as well as the background.

In[3]:= **ShowWalk2D[Walk2D[10^3],**
**   DefaultColor -> Hue[.75],**
**   Background -> GrayLevel[0.7]]**

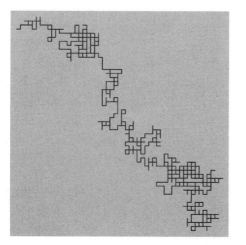

Out[3]= -Graphics-

As the graphics indicate, a lattice walk repeatedly revisits sites that have been previously visited in the course of its meandering. As a result, it is difficult, and usually impossible, to discern the history of the walk from a snapshot of the path. The best way to see the entire evolution process of the walk in an unobscured fashion is to create an animation.

ANIMATION

Animations (or movies) consist of a set of graphics cells shown in rapid succession. This can be done from within *Mathematica* using special features of the front end, or the graphics can be exported to special utilities such as QuickTime. In this section we will describe how to create a series of graphics using *Mathematica*. You should consult your User's Guide for platform-specific information.

We will create an animation using, as an example, an 8-step two-dimensional lattice walk.

```
In[1]:= w8 = Walk2D[8]

Out[1]= {{0, 0}, {0, -1}, {0, 0}, {0, -1}, {-1, -1},
          {-1, 0}, {-2, 0}, {-2, 1}, {-3, 1}}
```

Creating an animation in *Mathematica* is straightforward. The animation consists of a sequence of graphics cells where the first cell shows the first step (consisting of a line drawn between the first two elements in w8 for example) and each succeeding cell shows one more step than the previous cell. In general then, the *m*th cell is drawn using the Line function and the first $m + 1$ elements in w8. The graphics cells can be drawn using the following code.

```
In[2]:= Map[(Show[Graphics[Line[Take[w8, #]],
                DisplayFunction -> Identity]])&,
            Range[2, Length[w8]]];
```

The easiest way to see all of the cells of the animation simultaneously is to take the animation as an argument of the GraphicsArray function, using Partition to specify the number of cells in each line of the resulting graphics. Note the use of DisplayFunction to turn off the display of each graphic to the screen; GraphicsArray automatically overrides this value, so this construction only shows the entire array.

```
In[3]:= Show[GraphicsArray[Partition[%, 4]]]
```

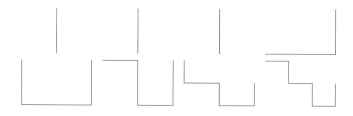

```
Out[3]= -GraphicsArray-
```

In general, objects in a graphics cell are scaled to fill the monitor screen. Therefore, if we simply create cells, each containing a different number of steps of the walk using the above graphics command, steps in one cell will appear to be of a different length then the same steps in other cells. This will result in a jerky-looking animation.

We can use the `PlotRange` option to make all of the step lengths in all of the graphics cells uniform. The ordered pair of the minimum and maximum values of the components of the random walk in each direction, $\{\{x_{min}, x_{max}\}, \{y_{min}, y_{max}\}\}$ can be determined by separating the x- and y-components of the walk using `Transpose` and then mapping an anonymous function containing `Min` and `Max` onto it.

```
In[4]:= Map[{Min[#], Max[#]}&, Transpose[w8]]

Out[4]= {{-3, 0}, {-1, 1}}
```

Here, then, is the command for creating the animation using `PlotRange` to control the size of each cell produced. In addition, we have added a red ball that moves to the current position in the walk which helps visualizing the animation, since often the walk crosses back upon itself. It is not difficult to change this marker to any desired shape or color.

```
In[5]:= AnimateWalk2D[coords_, opts___]:=
          Map[Show[Graphics[{
                  {RGBColor[1,0,0], PointSize[.02],
                   Point[ coords[[#]] ]},
                  Line[Take[coords, #]]}],
             opts,
             AspectRatio -> Automatic,
             PlotRange -> Map[{Min[#]-.2, Max[#]+.2}&,
                             Transpose[coords]]]&,
                         Range[2, Length[coords]]]
```

Note: We have added 0.2 to the maximum x and y values and subtracted 0.2 from the minimum x and y values in order to enhance the display by making the graphics a little smaller inside of its bounding box.

```
In[6]:= Show[GraphicsArray[Partition[
              AnimateWalk2D[Walk2D[10],
                  DisplayFunction -> Identity], 5]
          ]];
```

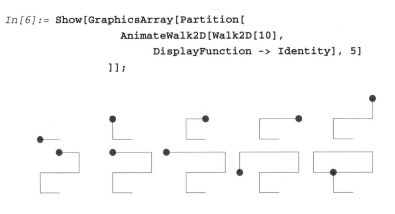

```
Out[6]= -Graphics-
```

Of course, when running the animation on a computer, you would just use `AnimateWalk2D` and not the `GraphicsArray` (examples are given on the CD-ROM). When the animation is run, the red ball can be seen moving around the screen landing on each successive step of the walk. This is especially striking when run on walks of length 100, or 1000, or more.

1.4 ■ COMPUTER SIMULATION PROJECTS

1. Two related quantities of interest in the random walk model are the mean number of distinct sites visited during a lattice walk (the number of sites that are visited at least one time) and the mean occupancy of a site (the mean number of times each site is visited). Determine the power law dependence of these quantities on the number of steps for the two-dimensional lattice walk.

2. Write a program for computing the step locations of a three-dimensional walk on a cubic lattice. Create an animation of a 10-step cubic lattice walk.

3. The *off-lattice* walk in two dimensions consists of n steps of equal length, each step being randomly oriented with respect to preceding steps. In this model, the x- and y-components of each step increment are not independent, because each step must have length 1. To satisfy this constraint, first randomly select an orientation angle `theta`, between 0 and 360 degrees, for each step. This can be accomplished using `Random[Real, {0, N[2 Pi]}]`. Then calculate the ordered pair consisting of the x- and y-components of the step, using `{Cos[theta], Sin[theta]}`. A list of the step increments in the off-lattice, n-step walk is given by first generating a list of `theta` values and then mapping the anonymous function `{Cos[#], Sin[#]}&` onto the list. Here is a program for doing this.

```
OffLatticeWalk[n_] :=
   FoldList[Plus, {0.0, 0.0},
            Map[{Cos[#],Sin[#]}&,
               Table[Random[Real,{0, N[2Pi]}], {n}]]]
```

Write a program for the step locations of an off-lattice walk in three dimensions.

REFERENCES

* George H. Weiss. Random walks and their applications. *American Scientist* 71 (1983) 65–71.

M. N. Barber and B. W. Ninham. *Random and Restricted Walks: Theory and Applications.* Gordon and Breach 1970.

F. Spitzer. *Principles of Random Walk*, Second Edition. Springer-Verlag Berlin 1976.

CHAPTER 2
The Self-Avoiding Walk

INTRODUCTION

The random walker described in Chapter 1 takes each step without any regard to previous steps. If a walker avoids visiting any location more than once, perhaps because he plants a landmine at each step location, the properties of the walk (*e.g.*, the mean dimensions) are fundamentally different than those of the random walk.

This self-avoiding walk (SAW) as it has come to be known, has several areas of application. The leading area is in polymer physics, where SAWs can represent long chain molecules consisting of many small molecules connected together by covalent chemical bonds (a polymer chain can be visualized as a pearl necklace). Other systems that use the SAW model include the zero-component ferromagnet, the $N \to 0$ limit of the N-vector model (a generalization of the Ising model), and critical phenomena in general.

The SAW is quite difficult to investigate analytically (with paper and pencil) and computational experimentation has proven to be an indispensable tool for studying the model. In this chapter we will look at the computation of the mean square, end-to-end distance of the SAW.

In studying the average properties of the random walk, a number of n-step walks are examined, each of which is grown by adding one step after another until n steps have been executed. This method of generating a sample of walks is not useful for studying the SAW because the occurrence of a single misstep, resulting from a step intersecting a previous step, makes it necessary to discard the walk and start a new one from scratch. The likelihood of such a misstep becomes increasingly likely as the walk proceeds and hence, the attrition rate for generating SAWs by growing them is prohibitively high.

An alternative approach to producing SAWs is to start with a self-avoiding walk and try to rearrange its shape, counting the result of each rearrangement attempt as another SAW. We will look at two well-known algorithms for accomplishing this rearrangement: the slithering snake algorithm and the pivot algorithm.

2.1 ■ THE SLITHERING SNAKE ALGORITHM

In the following description of the algorithm, the sequence of steps **2** through **4** is executed a number of times, first using the initial SAW configuration given in step **1**, and then using the SAW configuration resulting from the previous run-through of the sequence. We will describe the sequence in terms of an arbitrary SAW configuration, which we'll call `config`.

1. **A.** Create a self-avoiding walk of n steps on a two-dimensional square lattice where each step is in the same direction.

   B. Calculate the square end-to-end distance of the SAW and call it `squaredist`.
2. Randomly select a step increment and use it to add a step to `config`.
3. Check whether the new step intersects any previous step (excepting the first step) of `config`; that is, check whether the final location of the walk, after the new step, coincides with the location of another step in `config`. If it does, reverse `config` (*i.e.*, reverse the order of steps in `config`) and call the resulting SAW path. If the final location does not so coincide, add the new step to the end of `config`, discard the first step of `config` (thereby conserving the number of steps in the walk), and call the resulting SAW `path`.
4. Calculate the square end-to-end distance of `path`, add the value to `squaredist`, and return `path`.
5. Execute the sequence of steps **2** through **4** m times, starting with the initial SAW.
6. Determine the mean square, end-to-end distance.

We'll now implement the slithering snake algorithm in a step-by-step fashion.

IMPLEMENTATION OF THE SLITHERING SNAKE ALGORITHM

1. **A.** The initial SAW configuration is given by

```
config = Table[{i, 0}, {i, 0, n}]
```

   B. The square end-to-end distance of the initial SAW is given by

```
squaredist = n^2
```

2. A step increment is randomly chosen and added to the last element in `config` to produce a new step location.

```
newpt = {{1,0}, {-1,0},
         {0,1}, {0,-1}}[[Random[Integer, {1,4}]]] +
       Last[config]
```

3. newpt is checked against config to see if it coincides with any step of config, other than the first step.

```
MemberQ[Rest[config], newpt]
```

If there is an overlap, config is reversed and the resulting SAW configuration is called path.

```
path = Reverse[config]
```

If there is no overlap, the first step of config is removed, newpt is added to the result, and the new SAW configuration is called path.

```
path = Join[Rest[config], {newpt}]
```

These three fragments are brought together in a conditional function:

```
If[MemberQ[Rest[config], newpt],
  path = Reverse[config],
  path = Join[Rest[config], {newpt}]]
```

4. The square end-to-end distance of the new SAW configuration, path, resulting from step **3** is computed and added to squaredist.

```
squaredist += Apply[Plus, (First[path] - Last[path])^2]
```

The value of the new SAW is then returned using path.

We can combine steps **2** through **4** in an anonymous function, using the # symbol in place of config (we'll call the anonymous function snake for convenience).

```
snake = (Module[{choice = {{1, 0}, {-1, 0},
                           {0, 1}, {0, -1}}},
   newpt = Last[#] + choice[[Random[Integer, {1, 4}]]];
   If[MemberQ[Rest[#], newpt],
     path = Reverse[#],
     path = Join[Rest[config], {newpt}] ];
   squaredist += Apply[Plus, (First[path] - Last[path])^2];
   path])&
```

5. The repeated application of the sequence of steps **2** through **4** *m* times is performed using the Nest function with the anonymous snake function and the initial SAW configuration.

```
Nest[snake, Table[{i, 0}, {i, 0, n}], m]
```

6. The mean square, end-to-end distance of the SAW is given by:

```
N[squaredist/(m + 1)]
```

We can now put all of these code fragments together into a program.

THE SLITHERING SNAKE PROGRAM

```
In[1]:= SlitheringSAW[n_, m_] := Module[{squaredist,snake},
            squaredist = n^2;
            snake = (Module[{newpt, path,
                             choice = {{1,0}, {-1,0},
                                       {0,1}, {0,-1}}},
                  newpt = Last[#] + choice[[Random[Integer,{1,4}]]]];
                  If[MemberQ[Rest[#], newpt],
                     path = Reverse[#],
                     path = Join[Rest[#], {newpt}] ];
                  squaredist +=
                         Apply[Plus, (First[path] - Last[path])^2];
                  path])& ;

            Nest[snake, Table[{i, 0}, {i, 0, n}], m];
            N[squaredist/(m + 1)]
            ]
```

Looking at how the mean square end-to-end distance of a 10-step self-avoiding walk changes as a function of the number of slithering movements, we see that while a single rearrangement from the initial fully extended configuration (with a square end-to-end distance of 100) reduces the dimensions of the walk only slightly, there is a substantial change after many rearrangements by slithering.

```
In[2]:= {(SeedRandom[3]; SlitheringSAW[10, 1]),
         (SeedRandom[3]; SlitheringSAW[10, 100])}

Out[2]= {91., 24.7921}
```

The SlitheringSAW program can be used to determine the critical exponent of the mean square end-to-end distance and the mean square radius of gyration for walks generated by the slithering snake algorithm. This can be done using the numerical analysis discussed in Chapter 1.

However, there are two drawbacks to using the slithering snake algorithm. One problem is that it is possible for the SAW to find itself in a "double cul-de-sac" shape from which it cannot extricate itself. An example of this situation is shown below.

```
In[3]:= Show[Graphics[
            Line[{{0,0}, {0,1}, {1,1}, {1,0}, {1,-1},
                {0,-1}, {-1,-1}, {-2,-1}, {-2,0},
                {-2,1}, {-1,1}, {-1,0}}],
            AspectRatio->1]]
```

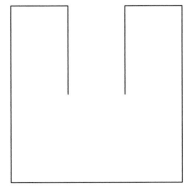

```
Out[3]= -Graphics-
```

Because a SAW can become trapped in a configuration from which it cannot escape, this SAW sampling method is *non-ergodic*.[1] (Nonetheless, the critical exponent obtained by the slithering snake algorithm has been found to be correct.)

Another problem with the slithering snake algorithm is that, as we showed above, because the walk moves one segment at a time, the SAW rearranges itself rather slowly, and a very large number of steps must be performed (*i.e.*, m must be large) in the simulation to obtain large-scale shape changes. Both of these shortcomings are addressed in the next SAW algorithm.

2.2 ▌ PIVOT ALGORITHM

The pivot algorithm is a very efficient dynamic algorithm for generating d-dimensional SAWs in a canonical ensemble (*i.e.*, with a fixed number of steps). It is based on randomly selecting one of the $2^d d!$ symmetry (rotation or reflection) operations on a d-dimensional lattice and applying the operation to the section of the SAW beyond a randomly selected step location. In two dimensions, it is sufficient (for ensuring ergodicity) to consider just three of the eight symmetry operations—specifically,

---

[1]A process is said to be *ergodic* if any sequence or significant sample
is equally representative of the whole.

rotations of +90, −90, and 180 degrees. To perform these rotations, three two-dimensional rotation matrices, rot90, rot180, and rot270 are defined.

```
In[1]:= rot90 = {{0,1}, {-1,0}}
Out[1]= {{0, 1}, {-1, 0}}


In[2]:= rot180 = {{-1,0}, {0,-1}}
Out[2]= {{-1, 0}, {0, -1}}


In[3]:= rot270 = {{0,-1}, {1,0}}
Out[3]= {{0, -1}, {1, 0}}
```

The use of these matrices can be illustrated by rotating the unit vector pointing in the east direction about the origin, $\{0, 0\}$.

```
In[4]:= initial = {{1,0}}
Out[4]= {{1, 0}}
```

Rotation is accomplished by matrix multiplication. The following list shows what happens to the initial point upon multiplication with each of the rotation matrices.

```
In[5]:= {initial.rot270, initial.rot90, initial.rot180}
Out[5]= {{{0, -1}}, {{0, 1}}, {{-1, 0}}}
```

The results show that the unit vector in the east direction is rotated to point in the south, north, and west direction by rot270, rot90, and rot180, respectively.

We can demonstrate the use of these rotation matrices to rearrange a two-dimensional SAW by pivoting a short chain.

```
In[6]:= chain = {{0,0}, {0,1}, {1,1}, {1,2}, {0,2}}
Out[6]= {{0, 0}, {0, 1}, {1, 1}, {1, 2}, {0, 2}}
```

The pivot procedure uses three quantities:

• the location pivot, of the pivot point along chain:

```
In[7]:= pivot = chain[[2]]
Out[7]= {0, 1}
```

- the coordinates `relative`, of the location of the *i*th step of `chain` relative to the pivot point:

```
relative = chain[[i]] - pivot
```

- the final coordinates move, of the *i*th step of `chain` after the pivot operation, where `roti` is `rot270`, `rot90`, or `rot180`:

```
move =  pivot + relative . roti
```

We now use these quantities to pivot the SAW `chain` about $\{0, 1\}$ in each of the three directions +90, −90, and 180 degrees.

```
In[8]:= Minus90RotSAW =
          {chain[[1]],
           chain[[2]],
           chain[[2]]  + (chain[[3]] - chain[[2]]). rot270,
           chain[[2]]  + (chain[[4]] - chain[[2]]). rot270,
           chain[[2]]  + (chain[[5]] - chain[[2]]). rot270}

Out[8]= {{0, 0}, {0, 1}, {0, 0}, {1, 0}, {1, 1}}


In[9]:= Plus90RotSAW =
          {chain[[1]],
           chain[[2]],
           chain[[2]]  + (chain[[3]] - chain[[2]]). rot90,
           chain[[2]]  + (chain[[4]] - chain[[2]]). rot90,
           chain[[2]]  + (chain[[5]] - chain[[2]]). rot90}

Out[9]= {{0, 0}, {0, 1}, {0, 2}, {-1, 2}, {-1, 1}}


In[10]:= Plus180RotSAW =
          {chain[[1]],
           chain[[2]],
           chain[[2]]  + (chain[[3]] - chain[[2]]). rot180,
           chain[[2]]  + (chain[[4]] - chain[[2]]). rot180,
           chain[[2]]  + (chain[[5]] - chain[[2]]). rot180}

Out[10]= {{0, 0}, {0, 1}, {-1, 1}, {-1, 0}, {0, 0}}
```

One of the advantages of programming in *Mathematica* is that we can use its graphics capabilities, not only to look at the results of a program, but also to debug the program as it is being developed. Here, we can visually check that the three operations performed above do in fact correctly rotate the SAW.

```
In[11]:= Show[GraphicsArray[Map[
         (Show[Graphics[
           {Line[#],
           {PointSize[0.08], RGBColor[1,0,0], Point[{0,1}]}},
           AspectRatio->1,PlotRange->{{-2,2}, {-1,3}}]])&,
         {chain, Plus90RotSAW, Minus90RotSAW, Plus180RotSAW}]]]
```

```
Out[11]= -GraphicsArray-
```

The first picture is the starting chain SAW configuration and the second, third, and fourth pictures are the SAW configurations that result from rotations of 90, −90, and 180 degrees about the second step location in `chain`. The pictures confirm that the pivot operation has been done correctly.

Building on the example above, we can write down the algorithm for performing the pivoting operation on any SAW. The sequence of steps **2** through **5** in the algorithm are executed a number of times, first using the initial SAW configuration given in step **1**, and then using the value of the SAW configuration resulting from the previous run-through of the sequence. We will describe the steps in terms of an arbitrary SAW configuration, which we'll call `config`.

1. A. Create an n-step SAW on a two-dimensional square lattice.
   B. Calculate the square end-to-end distance of the SAW and call it `squaredist`.
2. Choose at random, a pivot point k (where $0 < k < n$) along `config` and divide `config` into two parts: One part, consisting of the steps in `config` up to and including k, is called `fixsec`, and the other part, consisting of the steps in `config` subsequent to k, is called `movesec`.
3. Choose at random, a symmetry operation of the lattice.
4. Apply the operation to `movesec` to obtain the rotated chain section, and call it `newsec`.
5. Check if any of the step locations in `newsec` and `fixsec` coincide. If they do not coincide, create a new SAW configuration, naming it `newconfig`, by joining `newsec` and `fixsec`, calculate its square end-to-end distance and add that to `squaredist`, and return `newconfig`. If they do coincide, calculate the square end-to-end distance of the previous SAW configuration, `config`, add that to `squaredist`, and return `config`.

6. Execute the sequence of steps **2** through **5**, m times, starting with the initial SAW.

7. Determine the mean square end-to-end distance.

IMPLEMENTATION OF THE PIVOT ALGORITHM

1. A. The initial SAW configuration is given by

```
Table[{j, 0}, {j, 0, n}]
```

B. The square end-to-end distance of the initial SAW is given by

```
squaredist = n^2
```

2. A pivot point is randomly selected.

```
ball = Random[Integer, {1, n - 1}]
```

Using `ball`, `config` is divided into two sections, the first section comprising the steps in `config` up to and including the pivot point,

```
fixsec = Take[config, ball]
```

and the second section comprising the steps in `config` subsequent to the pivot point.

```
movesec = Take[config, ball - (n + 1)]
```

3. One of the three rotation matrices (corresponding to rotations of -90, $+90$ or 180 degrees) is randomly selected using the function `rotchoice`.

```
rotchoice = {{{0,-1}, {1,0}},
             {{0,1}, {-1,0}},
             {{-1,0}, {0,-1}}}[[ Random[Integer, {1, 3}] ]]
```

4. The step locations in `movesec` are rotated.

```
newsec = Map[Function[y, config[[ball]] +
             (y - config[[ball]]) . rotchoice], movesec]
```

Note: This computation is simply a generalization of the pivot operation we demonstrated in the short SAW example beginning on page 21.

5. `newsec` is checked against `fixsec` to see if any of the step locations coincide.

```
Intersection[newsec, fixsec] == {}
```

If there is no step intersection, `fixsec` and `newsec` are joined together to form a new SAW configuration, which is called `newconfig`. The square end-to-end distance is then calculated and added to `squaredist`, and the new SAW configuration is returned.

```
newconfig = Join[fixsec, newsec];
squaredist += Apply[Plus, Last[newconfig]^2];
newconfig
```

If there is an intersection of steps, the end-to-end distance of the previous SAW configuration, `config` is calculated and added to `squaredist`, and the previous SAW is returned.

```
squaredist += Apply[Plus, Last[config]^2];
config
```

All of the code fragments in step **5** are brought together in a conditional function.

```
If[Intersection[newsec, fixsec] == {},
    newconfig = Join[fixsec, newsec];
    squaredist += Apply[Plus, Last[newconfig]^2];
  newconfig,
    squaredist += Apply[Plus, Last[config]^2];
    config]
```

We can combine steps **2** through **5** in an anonymous function `twistAndShout` using the # symbol in place of `config`.

```
twistAndShout =
(Module[{ball, fixsec, movesec, rotchoice, newsec, newconfig,
        rot ={{{0,-1},{1,0}},{{0,1},{-1,0}},
              {{-1,0},{0,-1}}}},
   ball = Random[Integer, {1, n-1}];
   fixsec = Take[#, ball];
   movesec = Take[#, ball - (n + 1)];
   rotchoice = rot[[Random[Integer, {1, 3}] ]];
   newsec = Map[Function[y, #[[ball]] +
                (y - #[[ball]]) . rotchoice],
              movesec];
   If[Intersection[newsec, fixsec] == {},
       newconfig = Join[fixsec,newsec];
       squaredist += Apply[Plus, Last[newconfig]^2];
     newconfig,
       squaredist += Apply[Plus, Last[#]^2]; #]
])&
```

6. The repeated application (*m* times) of the sequence of steps **2** through **5** is performed using `Nest[twistAndShout, Table[j, 0, j, 0, n], m]`.

7. The mean square end-to-end distance is given by `N[squaredist/(m + 1)]`.

THE PIVOT PROGRAM

```
In[1]:= Pivot2DSAW[n_Integer, m_Integer] :=
          Module[{squaredist,twistAndShout},
              squaredist = n^2;
              twistAndShout =
              (Module[{ball, fixsec, movesec, rotchoice,
                      newsec, newconfig,
                      rot ={{{0,-1}, {1,0}},
                            {{0,1}, {-1,0}},
                            {{-1,0}, {0,-1}}}},
                  ball = Random[Integer, {1, n-1}];
                  fixsec = Take[#, ball];
                  movesec = Take[#, ball - (n + 1)];
                  rotchoice = rot[[ Random[Integer, {1, 3}] ]];
                  newsec = Map[Function[y,#[[ball]] +
                                        (y-#[[ball]]).rotchoice],
                              movesec];
                  If[Intersection[newsec, fixsec] == {},
                     newconfig = Join[fixsec, newsec];
                     squaredist += Apply[Plus, Last[newconfig]^2];
                     newconfig,
                     squaredist += Apply[Plus, Last[#]^2];
                     #]])&;
              Nest[twistAndShout, Table[{j, 0}, {j, 0, n}], m];
              N[squaredist/(m + 1)]]
```

The mean square end-to-end distance for a 10-step SAW, as a function of the number of pivots executed, can be calculated.

```
In[2]:= {(SeedRandom[3]; Pivot2DSAW[10,1]),
         (SeedRandom[3]; Pivot2DSAW[10,100])}

Out[2]= {76., 30.297}
```

As expected, a single pivot can produce a much bigger change in the SAW shape than can a single "slither," but for a large number of rearrangements, the two algorithms give, on average, the same results.

2.3 ■ COMPUTER SIMULATION PROJECTS

1. A property of the SAW that is of considerable interest is the so-called *fractal dimension* of the SAW. The fractal dimension d_f of a walk is given by the relationship $< r^2 >^{1/2} = n^{1/d_f}$. SAWs, like random walks, have a critical exponent v, given by the relationship $< r^2 >= n^v$ and it follows that $d_f = 2/v$ for SAWs or random walks. The dimensional dependence of d_f is different for the random walk and the SAW. The fractal dimension of a random walk is independent of the dimensionality of the walk; *i.e.*, a random walk executed on a d-dimensional lattice has $d_f = 2$, regardless of the value of d (you can create and run one- and three-dimensional versions of the lattice walk simulation program to confirm this). In contrast to the random walk, the fractal dimension of SAWs is dimensionally dependent. In one dimension, the end-to-end distance of the SAW must be proportional to the number of steps, because backtracking is not allowed. So, $< r^2 >= n^2$, and hence $d_f = 1$ and $v = 2$. Going to higher dimensions, v decreases, eventually approaching the random walk value of 1, because the chance of self-intersection decreases in higher dimensions.

Determine the value of d_f for a two-dimensional SAW on a square lattice.

Hint: For a d-dimensional SAW ($d \leq 4$), $< r^2 >$ can be shown to be roughly proportional to $n^{6/(d+2)}$, so that v and d_f scale approximately as $6/(d + 2)$ and $(d + 2)/3$, respectively. Note that the critical exponent of a four-dimensional SAW is the same as that of a random walk.

2. The slithering snake algorithm can be modified by restricting new steps to being at either +90 degrees or −90 degrees to the end of the walk (*i.e.*, the last step). This can be done using either one of the following functions:

```
newpt = Last[#] +
        Complement[choice, {Apply[Subtract, #[[{-1,-2}]]]},
                   {Apply[Subtract, #[[{-2,-1}]]]}
                  ][[Random[Integer, {1,2}]]]
```

```
newpt = Last[#] +
        DeleteCases[choice, Apply[Subtract,#[[{-1,-2}]]] |
                            Apply[Subtract,#[[{-2,-1}]]]
                   ][[Random[Integer, {1,2}]]]
```

Explain how these code fragments work and then incorporate them into the SlitheringSAW program and determine their effect on the critical exponent of the SAW.

Hints: A step at 180 degrees corresponds to a step increment equal to the preceding step increment. A step at 360 degrees corresponds to a step increment equal to the negative of the preceding step increment.

3. The lack of ergodicity does not hinder the usefulness of the slithering snake algorithm for studying dynamic changes in the shape of an SAW. Modify the `SlitheringSAW` program so that it returns a list of the configurations of the walk for *n* slither attempts and use that program to create an animation of the slithering process as a function of time.

Hint: Each time step corresponds to an application of the slither function.

4. Create an animation of the pivoting process. The resulting picture depicts the rearrangements of a long chain polymer molecule in dilute solution.

5. Extend the SAW pivot algorithm to three dimensions on a cubic lattice.

Hint: Ergodicity is satisfied by using five symmetry operations, corresponding to the following rotations: 90 degrees in the *x-y* plane, −90 degrees in the *x-y* plane, 180 degrees in the *x-y* plane, 90 degrees in the *x-z* plane, and −90 degrees in the *x-z* plane. (This is equivalent to rotating a line between adjacent lattice sites to its 5 nearest neighbor sites.)

Here is the rotation of 90 degrees in *x-y* plane:

```
rot1  = {{0, 1, 0}, {-1, 0, 0}, {0, 0, 1}}
```

This gives a rotation of −90 degrees in *x-y* plane:

```
rot2 = {{0, -1, 0}, {1, 0, 0}, {0, 0, 1}}
```

For a rotation of 180 degrees in *x-y* plane:

```
rot3 = {{-1, 0, 0}, {0, -1, 0}, {0, 0, 1}}
```

For a rotation of 90 degrees in *x-z* plane:

```
rot4 = {{0, 0, 1}, {0, 1, 0}, {-1, 0, 0}}
```

For a rotation of −90 degrees in *x-z* plane:

```
rot5 = {{0, 0, -1}, {0, 1, 0}, {1, 0, 0}}
```

REFERENCES

* Gordon Slade. Self-avoiding walks. *Mathematical Intelligencer* 16 (1994) 29–35.

G. F. Lawler. *Intersections of Random Walks.* Birkhäuser Boston 1991.

Neal Madras and Gordon Slade. *The Self-Avoiding Walk.* Birkhäuser Boston 1992.

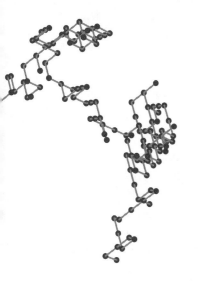

CHAPTER 3

Accretion

INTRODUCTION

Accretion is a process in which particles move around until they encounter another particle to "stick to." As particles coalesce, large, irregularly shaped clusters are formed. If the clusters are free-floating, the process is known as *aggregation*, and if the clusters are attached to a surface, the process is known as *deposition*.

In *diffusion-limited aggregation*, a particle undergoes Brownian motion until it makes contact with a free-floating cluster of particles which is known as a diffusion-limited aggregate (DLA). This mechanism underlies a wide variety of natural phenomena, including crystallization, colloidal and polymeric condensation, soot formation, and dielectric breakdown.

In *ballistic deposition*, a particle starts above a solid substrate or surface and follows a straight-line downward trajectory until it reaches the surface or makes contact with another particle. Deposition has many applications in the area of materials fabrication, such as thin-film formation, vapor deposition, sputtering, and molecular-beam epitaxy.

3.1 ◼ THE DIFFUSION-LIMITED AGGREGATION MODEL

The DLA model is most easily described in physical terms. At any time during DLA growth, the system consists of a cluster of particles, together with a particle executing a random walk. Initially, the cluster contains just a single seed particle. The cluster grows via a simple process: A particle starts at a randomly chosen location along the perimeter of a circle centered on the seed, and executes a random walk until the particle is either a certain distance outside the circle, in which case it vanishes, or until the particle is adjacent to the cluster, in which case it joins the cluster. This process is repeated, one particle at a time, until the cluster reaches a given size.

THE DLA ALGORITHM

The DLA model employs a two-dimensional square lattice. The sequence of steps **2** through **5** will be executed a number of times, first using the value of the initial cluster `occupiedSites`, given in step **1**, and then using the value of `occupiedSites` resulting from the previous run-through of the sequence.

1. Create a list, called `occupiedSites`, containing the lattice site $\{0, 0\}$.

2. Determine the lattice site nearest to a randomly chosen location along the circumference of a circle whose radius `rad` equals a specified value s, plus the maximum absolute coordinate value in `occupiedSites`.

3. Starting at the selected lattice site, execute a lattice walk until the step location is either at a distance greater than (`rad` + `s`), or on a site that is contiguous (adjacent) to a site in `occupiedSites`. Call the final step location of the walk `loc`.

4. If `loc` is adjacent to a site in the `occupiedSites` list, then add `loc` to `occupiedSites`.

5. Execute the sequence of steps **2** through **4** until the length of `occupiedSites` reaches a value n.

IMPLEMENTATION OF THE DLA ALGORITHM

1. Here is the list containing the seed site.

```
occupiedSites = {{0, 0}}
```

2. Here is the radius of the circle from which a particle starts its random walk.

```
rad = Max[Abs[occupiedSites]] + s
```

The walk starts from the lattice site nearest to a randomly chosen location on the circle.

```
Round[rad {Cos[#], Sin[#]}]&[Random[Real, {0, N[2 Pi]}]]
```

Each new step of the walk is generated using the following:

```
(# + {{1,0},{0,1},{-1,0},{0,-1}}[[Random[Integer, {1,4}]]])&
```

where # represents the current step location of the walk. (The one-dimensional analog to this anonymous function was discussed in Chapter 1.)

The random walk is terminated when the value `True` is returned by the following anonymous function.

```
(Apply[Plus, #^2] > (rad + s)^2 ||
        Intersection[occupiedSites,
                Map[Function[y, y + #],
                {{1,0}, {0,1}, {-1,0}, {0,-1}}]] != {} &)
```

The first test determines whether the step location is at a square distance greater than $(rad+s)^2$ from the seed site and the second test determines whether the step location is adjacent to a site in `occupiedSites`.

The final location `loc` of the random walk is determined using the `FixedPoint` function with the functions given above.

```
loc =
FixedPoint[
  (# + {{1,0},{0,1},
        {-1,0},{0,-1}}[[Random[Integer, {1,4}]]])&,
   Round[rad {Cos[#], Sin[#]}]&[Random[Real, {0, N[2 Pi]}]] ,
   SameTest -> (Apply[Plus, #2^2] > (rad + s)^2 ||
                Intersection[occupiedSites,
                        Map[Function[y, y + #2],
                        {{1,0},{0,1},{-1,0},{0,-1}}]] != {}
            &)]
```

Note: The `SameTest` option in `FixedPoint` uses #2 to refer to the last result generated.

3. `loc` is checked to see whether the walk ended because it was too far from the cluster; if not, it is added to `occupiedSites`.

```
If[Apply[Plus, loc^2] <= (rad +s)^2,
   occupiedSites = Join[occupiedSites, {loc}]]
```

4. The repeated application of the sequence of steps **2** through **3**, until the length of `occupiedSites` equals n, is performed with a conditional function.

```
While[Length[occupiedSites] < n,
   rad = (Max[Abs[occupiedSites]] + s);
   loc = FixedPoint[
        (# + {{1,0},{0,1},
              {-1,0},{0,-1}}[[Random[Integer, {1,4}]]])&,
         Round[rad {Cos[#], Sin[#]}]&[Random[Real,
                                         {0, N[2 Pi]}]],
         SameTest ->
            (Apply[Plus, #2^2] > (rad + s)^2 ||
                Intersection[occupiedSites,
                        Map[Function[y, y + #2],
                        {{1,0},{0,1},
                         {-1,0},{0,-1}}]] != {} &)];
   If[Apply[Plus, loc^2] <= (rad +s)^2,
      occupiedSites = Join[occupiedSites, {loc}]]]
```

We can now assemble these code fragments to create a program.

THE DLA PROGRAM

```
In[3]:= DLA[s_Integer, n_Integer] :=
        Module[{loc, rad, particleCount = 0,
              stepChoices = {{1,0},{0,1},{-1,0},{0,-1}}},
            occupiedSites = {{0,0}};
        While[Length[occupiedSites] < n,
            ++particleCount;
            rad = Max[Abs[occupiedSites]] + s;
            loc = FixedPoint[
               (# + stepChoices[[Random[Integer,{1,4}]]])&,
                Round[rad {Cos[#], Sin[#]}]&[Random[Real,
                                               {0,N[2Pi]}]],
                SameTest ->(Apply[Plus, #2^2] > (rad + s)^2 ||
                             Intersection[occupiedSites,
                               Map[Function[y, y + #2],
                                  stepChoices]] != {} &)
                ];
            If[Apply[Plus, loc^2] < rad^2,
               occupiedSites = Join[occupiedSites, {loc}]]
            ];
        Print["The number of particles released was ",
            particleCount];
        occupiedSites]
```

We have included a counter `particleCount`, to keep track of how many particles are released during the growth process and the final tally is printed out.

3.2 ■ VISUALIZING DIFFUSION-LIMITED AGGREGATION

The graphics of the DLA structure can be created using the `Graphics` primitive `Rectangle` to represent each particle in the cluster. The history of the accretion process can be seen by numbering and shading the rectangles in the order in which they were added to the cluster, with earlier rectangles being darker in color and lower in number.

A typical output from running the DLA program is shown below. The example shows a DLA that is allowed to grow to a size of 30 particles. So, in this example, it took the release of 109 meandering particles until 30 actually stuck to the structure (the others wandered outside of the specified radius).

```
In[4]:= ShowDLA[sites_,opts___]:= Module[{structure, s=Length[sites]},
            structure = {};
            Map[(AppendTo[structure,
                        {Hue[.99 #/s],
                          Rectangle[sites[[#]] - {0.5, 0.5},
                              sites[[#]] + {0.5, 0.5}],
                          RGBColor[1,1,1],
                          Text[#, sites[[#]]]}])&,
                Range[s]];
            Show[Graphics[structure],
                opts,
                Axes -> None,
                AspectRatio -> Automatic,
                PlotRange -> All]
            ]
```

```
In[5]:= agg = DLA[10, 30];
```

```
Out[5]= The number of particles released was  109
```

```
In[6]:= ShowDLA[ agg ]
```

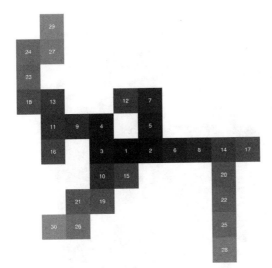

```
Out[6]= -Graphics-
```

For large DLAs, it is quicker to render each particle as a single point, as opposed to the above rectangles. Here is a program that accomplishes this and colors each point according to its history.

```
In[7]:= ShowDLA2[sites_, opts___Rule]:=
      Module[{len = Length[sites], points, colors},
            points = Map[Point, sites];
            colors = Map[Hue, Range[len]/len];
            Show[Graphics[{PointSize[1/Sqrt[len]],
                        Transpose[{colors,points}]},
                  opts, Axes -> None,
                  AspectRatio -> Automatic,
                  PlotRange -> All]]]
```

This function allows you to specify any number of additional graphics options to the plot. In addition, it computes a pointsize that scales with the size of the DLA. So, for example, you could display a DLA agg on a black background and connect the points with lines, by using ShowDLA2[agg, Epilog -> Line[agg], Background -> GrayLevel[0]].

Here, then, is a much larger DLA:

```
In[8]:= agg = DLA[2, 1000];
```

```
In[9]:= ShowDLA2[agg,
            Background -> GrayLevel[0]]
```

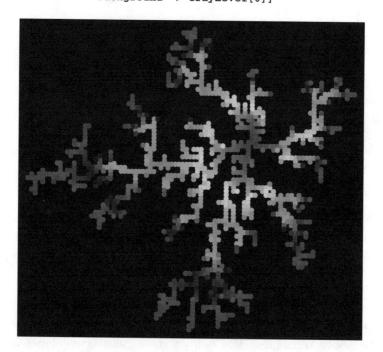

Out[9]= -Graphics-

3.3 ■ THE FRACTAL DIMENSION OF A DLA

As DLA growth proceeds, the shape of the aggregate or cluster becomes increasingly irregular and tenuous. This occurs as the result of a screening effect which increases the likelihood that a meandering particle will contact an exposed exterior portion of the DLA before it penetrates into a more shielded interior portion. (This effect can be seen by looking at the locations of sites in the cluster as a function of when they joined the cluster.)

We can get a feel for the compactness of a DLA (how it fills space) by measuring its fractal dimension. There are a number of fractal dimensions that can be measured (*e.g.*, the dependence of the radius of gyration of the cluster on the size of the cluster) and, while the various fractal measures give different quantitative results, they display a universal trend: As a DLA increases in size, its fractal dimension decreases. Here we will follow the density of space occupied by the cluster as a function of size.

It should be noted that the overall DLA process is *stochastic*, and it is therefore necessary to take an average of the fractal dimension (or any other quantity) over a number of randomly generated DLAs. However, for illustrative purposes, we will consider only a single DLA.

COMPUTING THE FRACTAL DIMENSION

The following steps are used to compute the fractal dimension of the DLA:

1. A `fractalDataList` of ordered pairs is constructed,

```
fractalDataList =
  Table[{2r, occSiteDensity[r]},
     {r, Max[Abs[occupiedSites]]}]
```

where `occupiedSites` is a list of the sites in the cluster and `occSiteDensity` is the number of sites in `occupiedSites` that lie within a square running between $-r$ and r in each direction, divided by the total number of sites in the square.

The value of `occSiteDensity` is calculated for a given square size as follows:

```
occSiteDensity[t_Integer] :=
   N[Count[occupiedSites, {x_?(Abs[#] <= t &),
                           y_?(Abs[#] <= t &)}]]/(2t +1)^2
```

2. The fractal dimension of the DLA structure is determined using

```
fractaldim = Fit[N[Log[fractalDataList]], {1, x}, x]
```

These computations are put together in a program `FractalDimension` in the next section.

THE FRACTAL DIMENSION PROGRAM

```
In[1]:= FractalDimension[occupiedSites_List] :=
        Module[{occSiteDensity, fractalDataList, fractaldim},
          occSiteDensity[t_Integer] :=
            N[Count[occupiedSites, {x_?(Abs[#] <= t &),
                    y_?(Abs[#] <= t &)}]]/(2t +1)^2;

          fractalDataList = Table[{2s, occSiteDensity[s]},
                              {s, Max[Abs[occupiedSites]]}];
          fractaldim = Fit[N[Log[fractalDataList]],
                          {1, x}, x];
          Print["The fractal dimension of the DLA is  ",
            Coefficient[fractaldim, x]];
        ]
```

The fractal dimension of a small DLA is calculated below.

```
In[2]:= FractalDimension[DLA[10, 12]]

The number of particles released was 46
The fractal dimension of the DLA is -0.748013
```

The negative slope obtained for the fractal dimension indicates that the density of the DLA structure is greatest at its center and decreases towards its periphery (see project 3 for a further discussion of this).

3.4 ■ THE BALLISTIC DEPOSITION MODEL

The ballistic deposition model is easily described in physical terms. The surface is initially smooth, consisting of a single row of particles. A particle is released at a randomly chosen location a certain distance above the surface. It falls straight downwards towards the surface until it reaches the surface or is adjacent to another particle, where it remains. The process is repeated until a specified number of particles have been released.

THE BALLISTIC DEPOSITION ALGORITHM

The model employs a two-dimensional rectangular lattice. The sequence of steps **2** through **6** will be executed a number of times.

1. Create a $2 \times n$ matrix in which the top row consists entirely of 0s and the bottom row consists entirely of 1s.

2. Randomly select a site, `depositColumn`, in the first row of the matrix; this value indicates which lattice column the particle will travel down.

3. Using the value of `depositColumn`, create the list `nnColumns` of the selected matrix column and its nearest neighbor columns to the right and left.
   When the selected column is the rightmost (leftmost) column, the nearest neighbor column on the right (left) is taken to be the leftmost (rightmost) column. (This is known as a *periodic boundary condition.*)

4. Find the positions of the first occupied site in each of the three columns in `nnColumns` and determine which one occurs first. Calculate the position `deposit-Row` in the selected column which is adjacent to the position which occurs earliest. (This value is the lattice row where the particle stops.)

5. Place the particle in the lattice in the position given by {`depositRow`, `deposit-Column`}.

6. Check that all the entries in the first row of the matrix are 0s (they will be unless the particle was placed there in the previous step); if not, then add a row of 0s to the top of the matrix.

7. Execute the sequence of steps **2** through **6** repeatedly, until the number of occupied sites is t.

Implementation of the Ballistic Deposition Algorithm

1. The initial lattice is created as a table of 0s and 1s.

   ```
   init = Transpose[Table[{0, 1},{n}]]
   ```

   In the following steps **2** through **6**, the # symbol represents the lattice configuration at a given time step.

2. This gives the location of the column in the lattice in which the particle falls:

   ```
   depositColumn = Random[Integer, {1, n}]
   ```

3. The selected column and its two nearest-neighbor columns are determined with the following anonymous function:

   ```
   nnColumns =
   Transpose[#][[{depositColumn - 1/. 0 -> n,
                  depositColumn,
                  depositColumn + 1 /. (n + 1) -> 1}]]&
   ```

4. The row position in the lattice where the falling particle stops is determined:

```
depositRow = Min[Flatten[Map[Function[y,
    First[Position[y, 1]]], nnColumns[#]]] - {0, 1, 0}]&
```

5. The particle is placed into its final resting place in the lattice using:

```
ReplacePart[#, 1, {depositRow[#], depositColumn}]&
```

We can combine steps **2** through **5** in an anonymous function newLat,

```
newLat =
(depositColumn = Random[Integer, {1, n}];

 nnColumns =
    Transpose[#][[{depositColumn - 1 /. 0 -> n,
                   depositColumn,
                   depositColumn + 1 /. (n + 1) -> 1}]]&;

 depositRow =
    Min[Flatten[Map[Function[y, First[Position[y, 1]]],
       nnColumns[#]]] - {0, 1, 0}]&;

 ReplacePart[#, 1, {depositRow[#], depositColumn}]
)&
```

where # represents the lattice.

6. A row of 0s is added to the top of the lattice resulting from applying newLat to the lattice, unless the topmost row is already empty.

```
emitLayer = If[#[[1]] != Table[0, {n}],
    Prepend[#, Table[0, {n}]], #]&
```

Here, # represents the result of applying newLat to the lattice.

7. The sequence of steps **2** through **6** is repeated t times, using Nest:

```
deposition = Nest[emitLayer[newLat[#]]&, init, t]
```

The program for ballistic deposition is given by combining the code fragments.

THE BALLISTIC DEPOSITION PROGRAM

```
In[1]:= MolecularDeposition[n_, t_] :=
          Module[{init, newLat, nnColumns,
                  depositRow, emitLayer},
             init = Transpose[Table[{0, 1},{n}]];

             newLat =
               (depositColumn = Random[Integer, {1, n}];
                nnColumns =
                  Transpose[#][[{depositColumn - 1 /.
                                  0 -> n,
                                  depositColumn,
                                  depositColumn + 1 /.
                                    (n + 1) -> 1}]]&;
                depositRow =
                  Min[Flatten[Map[Function[y,First[Position[y,1]]],
                       nnColumns[#]]] - {0, 1, 0}]&;
                ReplacePart[#, 1, {depositRow[#],
                          depositColumn}])&;

             emitLayer = If[#[[1]] != Table[0, {n}],
                          Prepend[#, Table[0, {n}]], #]&;

             Nest[emitLayer[newLat[#]]&, init, t]
             ]
```

3.5 ▪ VISUALIZING BALLISTIC DEPOSITION

A simple graphics display of the final state of a ballistic deposition process can be created using

```
In[2]:= ShowDeposition[sites_, opts___] :=
          ListDensityPlot[Reverse[sites /. {0 -> 1, 1 -> 0}],
            opts,
            AspectRatio -> Automatic,
            Mesh -> False,
            FrameTicks -> None,
            Frame -> None]
```

A typical graphical output of running the ballistic deposition program is shown below, starting with a row of 200 particles and releasing a total of 10,000 particles.

In[3]:= **ShowDeposition[MolecularDeposition[200, 10000]]**

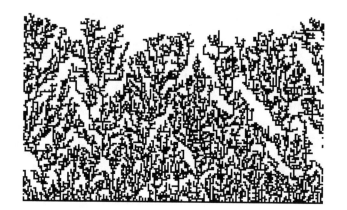

Out[3]= *-DensityGraphics-*

3.6 ■ COMPUTER SIMULATION PROJECTS

1. In computing the fractal dimension of the DLA, we calculated the density of occupied sites within a square box, as a function of the size of the box. Since a DLA structure becomes more tenuous (less space-filling) as it grows, due to the screening effect, the final DLA structure is more compact at its center than at its periphery. Thus, the density of a DLA decreases with increasing box size, in agreement with the negative slope obtained earlier.

An alternative measure of the fractal dimension can be made using the number of occupied sites within a square. Modify the `FractalDimension` program to compute the fractal dimension using the *mass* rather than the *density* of the DLA. Generate a DLA and compute both its mass and density fractal dimensions.

Hints: The mass fractal dimension value should be positive since the number of occupied sites increases with increasing box size. The value obtained should also be less than 2. Why is the value 2 expected for the mass fractal dimension of a completely compact (space-filling) two-dimensional structure?

2. It is interesting to watch the DLA growth process. This can be done by creating an animation using the `occupiedSites` list created with the `DLA` program in the following program. (The next-to-last two lines can be commented out when running the animation. They are included for the purposes of displaying the animation cells in an array.)

```
In[1]:= AnimateDLA[sites_List] := Module[{dynamics = {}},
           movies :=
              (AppendTo[dynamics,
                 Show[Graphics[
                    {RGBColor[0.9, 0.5, 0.9],
                     PointSize[0.115],
                     Point[sites[[#]]]},
                    Axes -> None,
                    PlotRange ->
                       Map[Function[y, {Min[y]-1, Max[y]+1}],
                          Transpose[sites]],
                    AspectRatio -> 1,
                    DisplayFunction -> Identity]]];
                 Show[dynamics, DisplayFunction -> $DisplayFunction])&;

           Show[GraphicsArray[Partition[
              Map[movies, Range[Length[sites]]], 6]]]]


In[2]:= AnimateDLA[DLA[10, 30]];
```

Create a DLA particle and make an animation of its development.

3. Create a three-dimensional version of the DLA program, in which a particle is released from a random location on the surface of a sphere, and executes a three-dimensional lattice walk until it touches a cluster or is a certain distance outside of the sphere. Write a program to determine the fractal dimension of the three-dimensional DLA.

Hints: Projects 2 and 3 in Chapter 1 deal respectively, with choosing a random location on a spherical surface and executing a walk on a cubic lattice.

A graphical image of a three-dimensional DLA can be created using the following program:

```
ShowDLA3D[sites_]:=
   Module[{structure, s = Length[sites]},
      structure = {};
      Map[(AppendTo[structure,
         {RGBColor[0.1, 0.9 #/s, 0.52],
          Icosahedron[sites[[#]],.375], RGBColor[1, 1, 1]}])&,
            Range[s]];
      Show[Graphics3D[structure],
         Axes -> False, Boxed -> False,
         AspectRatio -> Automatic, PlotRange -> All]
   ]
```

Create your own three-dimensional graphics using other graphic primitives.

4. The DLA model can be used to study the growth of crystals in materials. A multiple-nucleation growth process can be modeled by placing a number of "seeds" randomly throughout the lattice and keeping separate lists of the sites joining each cluster containing one of the seeds. Implement this variant of the DLA program.

5. Write a program that computes the fractal dimension of a ballistic deposition process.

6. Create a graphics display of the ballistic deposition process which uses coloring to indicate the history of the deposition process.

REFERENCES

* Leonard Sander. Fractal growth phenomena. *Scientific American* 256 (1987) 94–100.

* Brian Hayes. Nature's algorithm. *American Scientist* 82 (May/June 1994) 206–210.

Armin Bunde and Shlomo Havlin (eds.). *Fractals and Disordered Systems.* Springer-Verlag Heidelberg 1991.

Heinz-Otto Peitgen, Hartmut Jürgen, and Dietmar Saupe. *Chaos and Fractals: New Frontiers of Science.* Springer-Verlag 1992.

H. Eugene Stanley and Paul Meakin. Multifractal phenomena in physics and chemistry. *Nature* 335 (1988) 405–409.

Tamás Vicsek. *Fractal Growth Phenomena*, Second Edition. World Scientific Publishers Singapore 1992.

CHAPTER 4

Spreading Phenomena

INTRODUCTION

Recall the time you were sitting at your computer and you accidentally knocked over the cup of coffee or can of soda that you had placed within convenient reach. The flow of the beverage is a vivid, indeed horrific, example of the phenomenon of spreading. Spreading is a process in which an object extends itself over an increasingly larger area by incorporating regions adjacent to itself. A wide variety of natural processes can be described by the spreading or kinetic growth (KG) model. Examples include tumor growth, epidemic spread, gelation, rumor-mongering, and fluid flow through porous media.

The original KG model was the *Eden model*, which was introduced by a biologist (for whom it is named) to represent the growth of tumors. The Eden model operates on a two-dimensional square lattice system. Starting with a cluster list consisting of a seed site located at the origin, a site is randomly chosen from a perimeter list consisting of the nearest neighbor sites adjacent to the seed site (*i.e.*, the sites above [north], to the right [east], below [south], and to the left [west] of the seed site). The selected site is removed from the perimeter list and then placed in the cluster list. The nearest neighbor sites to the selected site are determined and those nearest neighbor sites that are not already in either the cluster list or the perimeter list are added to the perimeter list. Another site is then randomly selected from the perimeter list, and so on. This process continues until the cluster list reaches a certain size *n*.

We will look at two variations of the Eden model: The single percolation cluster model and the invasion percolation model.

4.1 THE SINGLE PERCOLATION CLUSTER MODEL

The single (or random) percolation cluster model describes the epidemic spread of disease. In this model, each randomly selected perimeter site has a probability p of joining the cluster. If the selected site is placed in the cluster, it is removed from the perimeter list. Even if the selected site is not placed in the cluster, it is removed from the perimeter list so that it cannot be chosen again later. When $p = 1$, this model reduces to the Eden model.

THE SINGLE PERCOLATION CLUSTER ALGORITHM

The single percolation cluster model takes place on a two-dimensional square lattice system. The sequence of steps **2** through **3** will be executed repeatedly, first using the values of the `cluster` and `perimeter` lists for the `seed`, given in step **1**, and then using the values of these lists resulting from the previous run-through of the sequence.

1. Create an ordered pair, consisting of a `cluster` list containing the `seed` site and a `perimeter` list of the nearest neighbor sites to the seed site, `{{{0, 0}}, {{1, 0}, {0, 1}, {-1, 0}, {0, -1}}}`. Also create an empty list called `reject`.
2. Randomly choose a site in the `perimeter` list.
3. Generate a random number and compare it to the input value p. If the random number is less than or equal to p, follow steps **3.A.i–3.A.ii**; if the random number is greater than p, follow steps **3.B.i–3.B.ii**.

    A. i. Determine the nearest neighbor sites to the selected site and put those nearest neighbor sites that are not already in the `perimeter` list, the `cluster` list, or the `reject` list into the `perimeter` list. Also, remove the selected site from the `perimeter` list.
    ii. Place the selected site in the `cluster` list and compute the ordered pair consisting of the new `cluster` list and the new `perimeter` list.
    B. i. Place the selected site in the `reject` list.
    ii. Remove the selected site from the `perimeter` list and compute the ordered pair consisting of the old `cluster` list and the new `perimeter` list.

4. Execute the sequence of steps **2** through **3** until the length of the `cluster` list reaches a value n, or until the `perimeter` list becomes empty.

IMPLEMENTATION OF THE SINGLE PERCOLATION CLUSTER ALGORITHM

1. The initial lists `{{seed site}, {nns to seed}}` and the `reject` list are written first.

```
{ {{0, 0}}, {{1,0}, {0,1}, {-1,0}, {0,-1}} };
reject = {}
```

The code for the sequence of steps **2** through **3** is written using an arbitrary ordered pair called `perClus`, where `perClus[[1]]` is the `cluster` list and `perClus[[2]]` is the `perimeter` list.

2. A site `select` in the `perimeter` list is randomly selected.

```
select = perClus[[2, Random[Integer,
                          {1, Length[perClus[[2]]]}]
               ]]
```

3. A random number is generated and checked against the value of p.

```
Random[] <= p
```

A. i. If the random number is less than or equal to p, `select` is removed from the `perimeter` list, and the nearest neighbors to `select` that are not already in `perClus[[1]]`, `perClus[[2]]`, or `reject` are added to the `perimeter` list `newPers`.

```
newPers =
   Complement[Union[Map[(# + select)&,
                        {{1,0},{0,1},{-1,0},{0,-1}}],
                   perClus[[2]]],
              {select}, perClus[[1]],reject]
```

B. ii. `select` is placed in `perClus[[1]]` and the ordered pair consisting of the new cluster list and the new perimeter list is created.

```
{Join[perClus[[1]], {select}], newPers}
```

C. i. If the random number is greater than p, `select` is placed in `reject`.

```
reject = Join[reject, {select}]
```

D. i. `select` is removed from the `perimeter` list and the ordered pair consisting of the unchanged `cluster` list and the new `perimeter` list is created.

```
{perClus[[1]], Complement[perClus[[2]], {select}]}
```

Overall, step **3** is expressed using the quantities given above in a conditional function.

```
If[Random[] <= p,
  newPers =
    Complement[Union[Map[(# + select)&,
                         {{1,0},{0,1},{-1,0},{0,-1}}],
                     perClus[[2]]],
              {select}, perClus[[1]], reject];
    {Join[perClus[[1]],{select}], newPers},
    reject = Join[reject, {select}];
    {perClus[[1]], Complement[perClus[[2]], {select}]}]
```

We can combine steps **2** through **3** in an anonymous function `pickAndChoose`, using the # symbol to represent `perClus`.

```
pickAndChoose :=
  (select = #[[2, Random[Integer, {1,Length[#[[2]]]}]]];

  If[Random[] <= p,
     newPers =
       Complement[Union[ Map[(# + select)&,
                         {{1,0}, {0,1}, {-1,0}, {0,-1}}],
                     #[[2]]],
                   {select},
                   #[[1]],
                   reject];
        {Join[#[[1]],{select}], newPers},
       reject = Join[reject, {select}];
       {#[[1]], Complement[#[[2]], {select}]}])&
```

4. The repeated application of the sequences of steps **2** through **3** on the initial value of `perClus` until the `perimeter` list is empty or the `cluster` list reaches size n is accomplished with the built-in function `FixedPoint`.

```
FixedPoint[pickAndChoose,
           {{{0,0}}, {{1,0}, {0,1}, {-1,0}, {0,-1}}},
           SameTest -> (#2[[2]] == {} ||
                        Length[#2[[1]]] == n &)]
```

Finally, the `cluster` list is returned by taking the first part of the above result.

Now we can combine these fragments into the single percolation cluster program.

THE SINGLE PERCOLATION CLUSTER PROGRAM

```
In[1]:= Epidemic[n_, p_]:=
          Module[{choices = {{1,0},{0,1},{-1,0},{0,-1}},
                  reject, pickAndChoose, select, newpers},
            reject = {};
            pickAndChoose :=
              (select = #[[2,Random[Integer,{1,Length[#[[2]]]}]]];
               If[Random[] <= p,
                 newPers =
                   Complement[Union[Map[Function[y, y + select],
                                        choices], #[[2]]],
                              {select}, #[[1]], reject];
                 {Join[#[[1]],{select}], newPers},
                 reject = Join[reject, {select}];
                 {#[[1]], Complement[#[[2]], {select}]}])&;
            FixedPoint[pickAndChoose, {{{0, 0}}, choices}, n,
                 SameTest -> (#2[[2]] == {} ||
                                Length[#2[[1]]] == n &)][[1]]
          ]
```

4.2 ■ INVASION PERCOLATION MODEL

The invasion percolation model describes the flow of fluid through porous media, such as occurs in tertiary oil recovery. In this model, each site in the perimeter list has a random number associated with it and the cluster spreads by incorporating the perimeter site with the lowest associated random number. Thus, this model describes a process of spreading that "follows the path of least resistance."

THE INVASION PERCOLATION ALGORITHM

The invasion percolation model takes place on a two-dimensional square lattice system. Steps **2** through **5** will be executed repeatedly, first using the nested list {cluster list, perimeter list} created in step **1**, and then using the values of these lists resulting from the previous run-through of the sequence.

1. First, create an ordered pair such that the first component is the cluster list containing the seed site, {{0, 0}}. The second component is the perimeter list consisting of ordered pairs whose first element is a randomly generated number and whose second element is a nearest neighbor site to the seed site.
2. Select the site in the perimeter list with the lowest attached random number.
3. Determine the nearest neighbor sites to the chosen site that are not already in either the cluster list or the perimeter list and pair them off with randomly generated numbers.

4. Remove the chosen site and its associated random number from the `perimeter` list and put the new nearest neighbor sites with their associated random numbers into the `perimeter` list.

5. Create an ordered pair consisting of the new `cluster` list with the chosen site, and the new `perimeter` list.

6. Execute the sequence of steps **2** through **5** repeatedly until the length of the cluster list reaches a specified value n.

IMPLEMENTATION OF THE INVASION PERCOLATION ALGORITHM

1. An ordered pair containing the `seed` site and the perimeter list is created as follows:

```
{{{0,0}},
 Transpose[{Table[Random[], {4}],
              {{1,0},{0,1},{-1,0},{0,-1}}}}]
```

In writing the code for the sequence of steps **2** through **5**, we will use `clusPer` where `clusPer[[1]]` is the `cluster` list and `clusPer[[2]]` is the `peri-meter` list.

2. The `perimeter` site with the lowest random number is found by sorting the second part of `clusPer` and taking the second component of the first ordered pair in the sorted list.

```
newcluSite = Sort[clusPer[[2]]][[1, 2]]
```

3. Here are the nearest neighbors to `newcluSite`.

```
nn = Map[(# + newcluSite)&, {{1,0},{0,1},{-1,0},{0,-1}}]
```

The sites in nn that are in either the `cluster` list or the `perimeter` list are removed from nn.

```
newnn = Complement[nn, clusPer[[1]],
            Transpose[clusPer[[2]]][[2]]]
```

The remaining sites are paired off with random numbers.

```
newpers =
  Transpose[{Table[Random[], {Length[newnn]}], newnn}]
```

4. newpers is added to the `perimeter` list, and `newcluSite`, with its associated random number, is removed from the perimeter list.

```
newPerLis =
    Join[DeleteCases[clusPer[[2]], {_, newcluSite}], newpers]
```

5. The ordered pair consisting of the new `cluster` list and the new `perimeter` list is formed.

```
{Join[clusPer[[1]], {newcluSite}], newPerLis}
```

We can combine steps **2** through **5** into an anonymous function `pickAndChoose`, using the # symbol to represent `clusPer`.

```
pickAndChoose :=
    (newcluSite = Sort[#[[2]]][[1, 2]];
    nearest neighbor =
        Map[Function[y, y + newcluSite],
            {{1,0},{0,1},{-1,0},{0,-1}}];
    newnn = Complement[nn, #[[1]],
                        Transpose[#[[2]]][[2]]];
    newpers = Transpose[{Table[Random[],
                                {Length[newnn]}], newnn}];
    newPerLis = Join[DeleteCases[#[[2]],
                        {_, newcluSite}], newpers];
    {Join[#[[1]], {newcluSite}], newPerLis})&
```

6. The sequence of steps **2** through **5** is repeatedly applied to the initial `clusPer` list until the `cluster` attains a size n, and the final list of `cluster` sites is then obtained.

```
cluster =
    Nest[pickAndChoose,
        {{0,0}, Transpose[{Table[Random[], {4}],
                            {{1,0},{0,1},{-1,0},{0,-1}}}]
        },
        n][[1]]
```

All of these code fragments are put into a complete program in the next section.

THE INVASION PERCOLATION PROGRAM

```
In[1]:= Invasion[n_Integer]:=
            Module[{pickAndChoose, nn, newnn, newpers, newPerLis,
                  choices = {{1,0}, {0,1}, {-1,0}, {0,-1}}},
              pickAndChoose :=
                  (newcluSite = Sort[#[[2]]][[1, 2]];
                  nn = Map[Function[y, y + newcluSite], choices];
                  newnn = Complement[nn, #[[1]],
                                      Transpose[#[[2]]][[2]]];
                  newpers = Transpose[{Table[Random[],
                                              {Length[newnn]}],
                                      newnn}];
                  newPerLis = Join[DeleteCases[#[[2]],
                              {_, newcluSite}], newpers];
                  {Join[#[[1]], {newcluSite}], newPerLis})&;

              Nest[pickAndChoose,
                  {{{0,0}}, Transpose[{Table[Random[], {4}],
                                      choices}]},
                  n][[1]]
            ]
```

4.3 ■ GRAPHICAL OUTPUT OF SPREADING MODELS

Pictures of the clusters created with any of the spreading programs can be generated using the same graphics commands—in this case the graphics primitive `Rectangle` is used.

```
In[1]:= ShowSpread[list_, opts___]:=
            Show[Graphics[{Hue[0.77],
                          Map[(Rectangle[# - {0.5,0.5},
                                  # + {0.5,0.5}])&, list]}],
                opts,
                AspectRatio -> 1,
                PlotRange -> Map[({Min[#], Max[#]})&,
                              Transpose[list]]]
```

Some typical spreading profiles are shown below. In the first example, an epidemic is shown that gradually spreads until 5000 individuals are infected. Here, the probability of each randomly selected perimeter site joining the cluster (becoming infected) is 1.0 (the Eden model).

In[2]:= **ShowSpread[Epidemic[5000, 1.0]]**

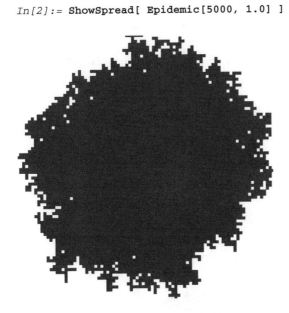

Out[2]= -Graphics-

This second example shows an epidemic that spreads to 5000 individuals, with each perimeter site having a probability of 0.58 of joining the cluster.

In[3]:= **ShowSpread[Epidemic[5000, 0.58]]**

Out[3]= -Graphics-

The following graphic shows an invasion stopped after 1000 time steps where sites are colored according to their history. This graphic was created with a slight modification to the ShowDLA2 program from the previous chapter. (Alternative graphics functions for displaying spreading graphics are included with the electronic distribution of files on the CD-ROM.)

4.4 ■ COMPUTER SIMULATION PROJECTS

1. Since spreading is inherently dynamic, an animation is appropriate (although it lacks the emotional intensity of a cinematographic presentation such as the 1958 movie classic *The Blob*). Create an animation of one of the spreading programs.

2. An interesting variant of the KG model incorporates the effect of obstacles on spreading. (This could serve, for example, as a model for fracture in a multi-component material.) Modify the invasion percolation model to include randomly placed sites that cannot be incorporated into the cluster.

3. If we are only interested in the $p = 1$ case, corresponding to the Eden model, then rather than running Epidemic[n, 1.0], it is preferable to write a program just for the $p = 1$ case by eliminating unnecessary computations from the Epidemic program. The resulting Eden program is given below.

```
Eden[n_] := Module[{choices = {{1,0},{0,1},{-1,0},{0,-1}},
                pickAndChoose,newPers, select, cluster},
    pickAndChoose :=
        (select = #[[2, Random[Integer,{1, Length[#[[2]]]}] ]];
         newPers = Complement[Union[Map[Function[y, y + select],
                                    choices], #[[2]]],
                            {select}, #[[1]]];
         {Join[#[[1]],{select}], newPers})&;

    cluster = Nest[pickAndChoose,
                {{{0,0}},choices}, n][[1]]
    ]
```

Note that it is not necessary to have a `reject` list or to use the conditional `If` function. In addition, the `FixedPoint` function can be replaced by `Nest`. Compare the relative speeds of `Epidemic[`n`, 1.0]` and `Eden[`n`]` for various values of n.

REFERENCES

* H. J. Hermann. Geometrical cluster growth models and kinetic gelation. *Physics Reports* 136 (1986) 153–227.

J. T. Chayes, L. Chayes, and C. M. Newman. The stochastic geometry of invasion percolation. *Communications in Mathematical Physics* 101 (1985) 383–407.

J. T. Chayes, L. Chayes, and C. M. Newman. Bernoulli percolation above threshold: an invasion percolation analysis. *Annals of Probability* 15 (1987) 1272–1287.

D. Wilkinson and J. F. Willemsen. Invasion percolation: a new form of percolation theory. *Journal of Physics A* 16 (1983) 3365–3376.

CHAPTER 5

Percolation Clustering

INTRODUCTION

When you make jello, leave milk out to curdle, or cut yourself and start bleeding, you are setting in motion the same process. The *sol-gel transition*, as it is called, consists of the conversion of a liquid to a semi-solid and it occurs as a result of the random joining together of molecules. This process, which is analogous to people in a crowd randomly joining hands with each other, is characteristic of percolation phenomenon.

Percolation is concerned with connectedness. P. G. de Gennes, winner of the 1991 Nobel Prize in Physics for his profound work on the theoretical physics of disordered materials, has described the percolation transition in the following way: "Many phenomena are made of random islands and in certain conditions, among these islands, one macroscopic continent emerges."

Percolation phenomena are widespread in nature. They occur in chemical systems (a polymerization reaction), biological systems (the immunological antibody-antigen reaction), and in physical systems (in critical phenomena). We will look at the clustering that occurs in the random site percolation model.

5.1 ▮ RANDOM SITE PERCOLATION

The random site percolation model consists of an $m \times m$ random Boolean lattice. This is a lattice in which the sites have values 0 and 1, where 0 represents an empty site and 1 represents an occupied site. The probability p of a site being occupied is independent of that of its neighbors. A cluster is defined as a group of occupied nearest-neighbor sites (nearest neighbor sites are those above, below, left, or right of a site).

The program can be described as follows:

- Each site on a two-dimensional $m \times m$ square lattice is assigned a random value (a = Random[]).
- If a <= p, its value is changed to 1; otherwise its value is changed to 0.
  This computation can be carried out in a single stroke using a one-liner program.

```
In[1]:= SitePercolation[p_, m_Integer] :=
            Table[Floor[1 + p - Random[]], {m}, {m}]
```

A typical run of the program produces

```
In[2]:= r = SitePercolation[0.35, 9]
```

```
Out[2]= {{0, 0, 1, 1, 0, 0, 0, 1, 0}, {0, 0, 1, 0, 0, 1, 0, 0, 0},
         {0, 0, 1, 0, 1, 1, 1, 0, 0}, {0, 0, 1, 1, 1, 1, 0, 0, 1},
         {1, 0, 0, 1, 1, 0, 0, 0, 0}, {0, 0, 1, 1, 1, 0, 0, 0, 0},
         {1, 0, 0, 0, 0, 1, 0, 1, 0}, {0, 1, 0, 0, 0, 1, 1, 1, 0},
         {0, 0, 0, 0, 0, 0, 0, 0, 0}}
```

It's useful to look at r in the form of a matrix.

```
In[3]:= r //MatrixForm
```

```
Out[3]//MatrixForm= 0   0   1   1   0   0   0   1   0
                    0   0   1   0   0   1   0   0   0
                    0   0   1   0   1   1   1   0   0
                    0   0   1   1   1   1   0   0   1
                    1   0   0   1   1   0   0   0   0
                    0   0   1   1   1   0   0   0   0
                    1   0   0   0   0   1   0   1   0
                    0   1   0   0   0   1   1   1   0
                    0   0   0   0   0   0   0   0   0
```

Looking at the MatrixForm of r, we can pick out seven clusters of connected non-zero sites. These clusters are easier to identify when presented as a graphic display.

While not necessary, it is helpful to visually match the positions of the elements in the MatrixForm of r with the associated graphics; *i.e.*, to have the element in the top-right corner of the matrix correspond to the top right square in the RasterArray graphic. This can be done by using Reverse[r] rather than r with RasterArray.

```
In[4]:= Show[Graphics[RasterArray[Reverse[r] /.
            {1 -> RGBColor[1,0,0],
             0 -> RGBColor[0,1,0]}]],
            AspectRatio -> 1]
```

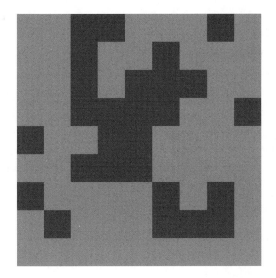

Out[4]= -Graphics-

We can also identify the clusters numerically, where each occupied lattice site has an integer value indicating the cluster to which it belongs. For example, this manually labels the clusters in `r`:

```
In[5]:= {{0, 0, 4, 4, 0, 0, 0, 7, 0}, {0, 0, 4, 0, 0, 4, 0, 0, 0},
        {0, 0, 4, 0, 4, 4, 4, 0, 0}, {0, 0, 4, 4, 4, 4, 0, 0, 6},
        {5, 0, 0, 4, 4, 0, 0, 0, 0}, {0, 0, 4, 4, 4, 0, 0, 0, 0},
        {3, 0, 0, 0, 0, 2, 0, 2, 0}, {0, 1, 0, 0, 0, 2, 2, 2, 0},
        {0, 0, 0, 0, 0, 0, 0, 0, 0}} //MatrixForm
```

Out[5]//MatrixForm=								
0	0	4	4	0	0	0	7	0
0	0	4	0	0	4	0	0	0
0	0	4	0	4	4	4	0	0
0	0	4	4	4	4	0	0	6
5	0	0	4	4	0	0	0	0
0	0	4	4	4	0	0	0	0
3	0	0	0	0	2	0	2	0
0	1	0	0	0	2	2	2	0
0	0	0	0	0	0	0	0	0

Once the clusters are identified numerically, there are a number of cluster-related quantities that are interesting to look at, such as the spatial characteristics of clusters (*e.g.*, their fractal dimensions) as a function of p and the percolation *threshold*, which is the value of p at which a spanning cluster (an uninterrupted path across the lattice) first appears. We will write a program to perform the labeling. We will use a well-known method, the Hoshen-Kopelman algorithm (named after its creators), which involves scanning r just one time.

This famous algorithm uses a Do loop. However, while Do is the most common programming construct for performing iterative computations in traditional procedural programming languages, it is usually *not* desirable to use the built-in Do function in a *Mathematica* program. This is because Do executes very slowly compared with other built-in *Mathematica* functions that perform iteration, such as Map, Fold, and Nest. An alternative (and faster) non-Do loop version of the cluster labeling program is given in project 2 of Chapter 13.

5.2 ■ CLUSTER LABELING

THE HOSHEN-KOPELMAN ALGORITHM

We'll be labeling a nested list r, consisting of m lists, each containing m elements which are either 0s or 1s.

1. A list u is created by adding: (a) a zero to the front of each of the lists in r and (b) a list of $(m + 1)$ zeros to the front of r (in the TableForm representation of r, this operation takes the $m \times m$ matrix and forms an $(m + 1) \times (m + 1)$ matrix by adding a top row of zeros and a left column of zeros).
2. A list ul is created which is the same size as u and consists of all zeros.
3. An empty list ulp is created.
4. The list u is scanned in a typewriter fashion, starting from position $\{2, 2\}$ and proceeding along each row in succession, going from the second element to the last element in the row. The elements in ul and ulp are changed during the scan of u according to the criteria in **A–F** below:

   Note: In the following step, we will refer to the elements in u and ul as *sites*. A site with a value of zero will be said to be empty and a site with a non-zero value will be said to be occupied. Additionally, we will refer to the site u[[i, j-1]], lying to the left of u[[i, j]], as uback, the site u[[i-1, j]], lying above u[[i, j]], as uup, the site ul[[i, j-1]], lying to the left of ul[[i, j]], as ulback, and the site ul[[i-1, j]], lying above ul[[i, j]], as ulup.

   A. At each site u[[i, j]], look to see whether it is occupied or empty and do the following:

   B. If u[[i, j]] is empty, go on to the next site.

   C. If u[[i, j]] is occupied, look at the nearest neighbor site in the previous column, uback, and the nearest neighbor site in the previous row, uup, and do the following:

D. If both uback and uup are empty, set ul[[i, j]] equal to one more than the current maximum value in ul and then add this new maximum value in ul to ulp.

E. If only one of the uback and uup sites is occupied, set ul[[i, j]] equal to the non-zero value of ulup or uback.

F. If both uback and uup are occupied, set ul[[i, j] equal to the smaller of ulp[[ulup]] and ulp[[ulback]] and set the value of the position in ulp having the larger of the values of ulp[[ulup]] and ulp[[ulback]] equal to the smaller of these values.

5. After the scan in step **4** is completed, relabel the occupied sites in ul so that connected sites (*i.e.*, occupied sites adjacent to occupied sites) have the same number.

6. Change the cluster numbers in ul so that they run sequentially, and without any gaps.

We can demonstrate how steps **4** through **5** work using a specific example. We'll work with the following matrix:

```
In[1]:= (r = {{1,0,0,1}, {1,0,1,1},
              {1,1,1,0}, {0,0,0,1}}) //MatrixForm

Out[1]//MatrixForm= 1   0   0   1
                    1   0   1   1
                    1   1   1   0
                    0   0   0   1
```

The lists u, ul, and ulp are given by:

```
In[2]:= (u = {{0,0,0,0,0}, {0,1,0,0,1}, {0,1,0,1,1},
              {0,1,1,1,0}, {0,0,0,0,1}}) //MatrixForm

Out[2]//MatrixForm= 0   0   0   0   0
                    0   1   0   0   1
                    0   1   0   1   1
                    0   1   1   1   0
                    0   0   0   0   1
```

```
In[3]:= (ul = u /. 1 -> 0) //MatrixForm
```

```
Out[3]//MatrixForm= 0   0   0   0   0
                    0   0   0   0   0
                    0   0   0   0   0
                    0   0   0   0   0
                    0   0   0   0   0
```

```
In[4]:= ulp = {}
```

```
Out[4]= {}
```

We now scan u according to step **4** in the algorithm. We'll state what changes we are making to ul and ulp and also refer to that part of step **4** on which the change is based.

1. At u[[2, 2]], we first change the value of ul[[2, 2]] to (1 + Max[ul]) and then add Max[ul] to ulp (step **4.D**).

```
In[5]:= ul[[2, 2]] = 1 + Max[ul];
        AppendTo[ulp, Max[ul]];
        {ul, ulp}
```

```
Out[5]= {{{0, 0, 0, 0, 0}, {0, 1, 0, 0, 0}, {0, 0, 0, 0, 0},
        {0, 0, 0, 0, 0}, {0, 0, 0, 0, 0}}, {1}}
```

2. At u[[2, 5]], we first change the value of ul[[2, 5]] to (1 + Max[ul]) and then add Max[ul] to ulp (step **4.D**).

```
In[6]:= ul[[2, 5]] = 1 + Max[ul];
        AppendTo[ulp, Max[ul]];
        {ul, ulp}
```

```
Out[6]= {{{0, 0, 0, 0, 0}, {0, 1, 0, 0, 2}, {0, 0, 0, 0, 0},
        {0, 0, 0, 0, 0}, {0, 0, 0, 0, 0}}, {1, 2}}
```

3. At u[[3, 2]], we change the value of ul[[3, 2]] to 1 (step **4.E**).

```
In[7]:= ul[[3, 2]] = 1;
        {ul, ulp}
```

```
Out[7]= {{{0, 0, 0, 0, 0}, {0, 1, 0, 0, 2}, {0, 1, 0, 0, 0},
        {0, 0, 0, 0, 0}, {0, 0, 0, 0, 0}}, {1, 2}}
```

4. At u[[3, 4]], we first change the value of ul[[3, 4]] to (1 + Max[ul]) and then add Max[ul] to ulp (step **4.D**).

```
In[8]:= ul[[3, 4]] = 1 + Max[ul];
        AppendTo[ulp, Max[ul]];
        {ul, ulp}
```

```
Out[8]= {{{0, 0, 0, 0, 0}, {0, 1, 0, 0, 2}, {0, 1, 0, 3, 0},
         {0, 0, 0, 0, 0}, {0, 0, 0, 0, 0}}, {1, 2, 3}}
```

5. A. At u[[3, 5]], the sites to the left of and above ul[[3, 5]] have values of 3 and 2 respectively. Therefore, we look at the values in the third and second elements in ulp and change the value of ul[[3, 5]] to the smaller of the ulp[[3]] and ulp[[2]] values, so that ul[[3, 5]] becomes 2 (step **4.F**).

```
In[9]:= ul[[3, 5]] = Min[ulp[[3]], ulp[[2]]];
        {ul, ulp}
```

```
Out[9]= {{{0, 0, 0, 0, 0}, {0, 1, 0, 0, 2}, {0, 1, 0, 3, 2},
         {0, 0, 0, 0, 0}, {0, 0, 0, 0, 0}}, {1, 2, 3}}
```

B. We also change the value of the position in ulp having the larger of the ulp[[3]] and ulp[[2]] values to the smaller of the ulp[[3]] and ulp[[2]] values, so that ulp[[3]] becomes 2 (step **4.F**).

```
In[10]:= ulp[[Max[ulp[[3]], ulp[[2]]]]] = Min[ulp[[3]], ulp[[2]]];
         {ul, ulp}
```

```
Out[10]= {{{0, 0, 0, 0, 0}, {0, 1, 0, 0, 2}, {0, 1, 0, 3, 2},
          {0, 0, 0, 0, 0}, {0, 0, 0, 0, 0}}, {1, 2, 2}}
```

6. At u[[4, 2]], we change the value of ul[[4, 2]] to 1 (step **4.E**).

```
In[11]:= ul[[4, 2]] = 1;
         {ul, ulp}
```

```
Out[11]= {{{0, 0, 0, 0, 0}, {0, 1, 0, 0, 2}, {0, 1, 0, 3, 2},
          {0, 1, 0, 0, 0}, {0, 0, 0, 0, 0}}, {1, 2, 2}}
```

7. At u[[4, 3]], we change the value of ul[[4, 3]] to 1 (step **4.E**).

```
In[12]:= ul[[4, 3]] = 1;
         {ul, ulp}
```

Out[12]= {{{0, 0, 0, 0, 0}, {0, 1, 0, 0, 2}, {0, 1, 0, 3, 2},
          {0, 1, 1, 0, 0}, {0, 0, 0, 0, 0}}, {1, 2, 2}}

8. At u[[4, 4]], we change the value of ul[[4, 4]] to the smaller of ulp[[1]] and ulp[[3]] and we change the value of the position in ulp having the larger of ulp[[1]] and ulp[[3]] to the smaller of ulp[[1]] and ulp[[3]]. Overall then, ul[[4, 4]] becomes 1 and ulp[[2]] becomes 1 (step **4.F**).

```
In[13]:= ul[[4, 4]] = Min[ulp[[1]], ulp[[3]]];
         ulp[[Max[ulp[[1]],
                 ulp[[3]]]]] = Min[ulp[[1]], ulp[[3]]];
         {ul, ulp}
```

Out[13]= {{{0, 0, 0, 0, 0}, {0, 1, 0, 0, 2}, {0, 1, 0, 3, 2},
          {0, 1, 1, 1, 0}, {0, 0, 0, 0, 0}}, {1, 1, 2}}

9. At u[[5, 5]], we first change the value of ul[[5, 5]] to (1 + Max[ul]) and then add Max[ul] to ulp (step **4.D**).

```
In[14]:= ul[[5, 5]] = 1 + Max[ul];
         AppendTo[ulp, Max[ul]];
         {ul, ulp}
```

Out[14]= {{{0, 0, 0, 0, 0}, {0, 1, 0, 0, 2}, {0, 1, 0, 3, 2},
          {0, 1, 1, 1, 0}, {0, 0, 0, 0, 4}}, {1, 1, 2, 4}}

This completes the scan of u. Here, now, is ul.

```
In[15]:= ul //TableForm
```

Out[15]//TableForm= 0	0	0	0	0
0	1	0	0	2
0	1	0	3	2
0	1	1	1	0
0	0	0	0	4

Notice that the sites identified in ul as belonging to clusters 3 and 2 are incorrectly labeled since all of the sites labeled 1, 2, or 3 actually belong to same cluster. Step **5** in the algorithm corrects the mislabeling.

ulp indicates the changes that need to be made in ul in order to correctly identify the clusters to which the sites belong. The way in which ulp works is as follows:

ulp[[k]] = m where m is an integer. If k = m, sites labeled as belonging to cluster k are correctly labeled. If m < k, sites labeled as belonging to cluster k need to be relabeled as belonging to cluster m. In our example,

```
In[16]:= ulp

Out[16]= {1, 1, 2, 4}
```

This output indicates that (a) the sites in ul that are labeled as belonging to cluster 1 actually belong to cluster 1; (b) the sites in ul labeled as belonging to cluster 2 actually belong to cluster 1; (c) the sites in ul labeled as belonging to cluster 3 actually belong to cluster 2; and (d) the sites in ul labeled as belonging to cluster 4 actually belong to cluster 4.

We can express the changes needed in ul as transformation rules.

```
In[17]:= Thread[Range[Length[ulp]] -> ulp]

Out[17]= {1 -> 1, 2 -> 1, 3 -> 2, 4 -> 4}
```

If we simply apply these transformation rules to ul, changing 2s to 1s and 3s to 2s, the resulting matrix will still be mislabeled since the site in ul previously labeled as belonging to cluster 3 will be relabeled as belonging to cluster 2, rather than cluster 1. To do the relabeling correctly, the transformation rules are applied to ul repeatedly.

```
In[18]:= (ul //. Thread[Range[Length[ulp]] -> ulp]) //MatrixForm

Out[18]//MatrixForm= 0  0  0  0  0
                     0  1  0  0  1
                     0  1  0  1  1
                     0  1  1  1  0
                     0  0  0  0  4
```

Now the sites belonging to the various clusters in our example are correctly identified.

Having worked through this simple example, we can now give the code to perform the steps in the algorithm. We'll refer to the $m \times m$ list that is input as r.

Step **1** is performed in the following two lines of code:

```
AddZeros[w_] := Prepend[w, Table[0, {Length[w[[1]]]}]]
u = Transpose[AddZeros[Transpose[AddZeros[r]]]];
```

Steps **2** and **3** are simply given by:

```
ul = u  /. 1 -> 0;
ulp = {}
```

Before implementing step **4**, we'll set up step **4.A**. This is expressed using the Which function with four successive tests, progressing from the most specific to the most general condition.

```
Which[u[[q, k]] == 1 && uup == 1 &&
                 uback == 1 && ulup =!= ulback,
        ul[[q, k]] = Min[ulp[[ulback]], ulp[[ulup]]];
        ulp[[Max[ulp[[ulback]],
                 ulp[[ulup]]]
          ]] = Min[ulp[[ulback]], ulp[[ulup]]],
    u[[q, k]] == 1 && uup == 1,
        ul[[q, k]] = ulup,
    u[[q, k]] == 1 && uback == 1,
        ul[[q, k]] = ulback,
    u[[q, k]] == 1,
        ul[[q, k]] = Max[ul] + 1;
        AppendTo[ulp, Max[ul]]
    ]
```

We can take the Which statement apart in order to see steps **4.D**, **4.E**, and **4.F** being implemented. The first test is the most specific.

```
u[[q, k]] == 1 && uup == 1 && uback == 1 && ulup =!= ulback
```

If it returns True, step **4.F** is carried out using:

```
ul[[q, k]] = Min[ulp[[ulback]], ulp[[ulup]]];
ulp[[Max[ulp[[ulback]], ulp[[ulup]]] ]] =
              Min[ulp[[ulback]], ulp[[ulup]]]
```

The second and third tests both refer to step **4.E** and are carried out if one of the tests returns True, after the first test fails.

```
u[[q, k]] == 1 && uup == 1, ul[[q, k]] = ulup;

u[[q, k]] == 1 && uback == 1, ul[[q, k]] = ulback
```

It is not necessary to consider the case of both adjacent sites in u being occupied and having the same value in ul, because in that case the second test is sufficient for making the necessary change in ul.

Here is the fourth and last test.

```
u[[q, k]] == 1
```

If this returns True after the other tests fail, step **4.D** is carried out using:

```
ul[[q, k]] = Max[ul] + 1;
AppendTo[ulp, Max[ul]]
```

We can now write step **4** by wrapping the Which statement in a Do loop.

```
Do[Which[u[[q, k]] == 1 && uup == 1 &&
                   uback == 1 && ulup =!= ulback,
         ul[[q, k]] = Min[ulp[[ulback]], ulp[[ulup]]];
         ulp[[ Max[ulp[[ulback]], ulp[[ulup]]] ]] =
           Min[ulp[[ulback]], ulp[[ulup]]],
      u[[q, k]] == 1 && uup == 1, ul[[q, k]] = ulup,
      u[[q, k]] == 1 && uback == 1, ul[[q, k]] = ulback,
      u[[q, k]] == 1, ul[[q, k]] = Max[ul] + 1;
      AppendTo[ulp, Max[ul]]],
  {q, 2, n+1}, {k, 2, m+1}]
```

Step **5** is carried out by first creating a list of integers running from 1 to the length of ulp,

```
new = Range[Length[ulp]]
```

and then using new with ulp to create a set of transformation rules:

```
relabelrules1 = Thread[new -> ulp]
```

We now create a list with the correct numbering (but possibly with gaps in the numbers) by repeatedly applying the transformation rules to ulp.

```
correctlabels = ulp //. relabelrules1
```

Step **6** is performed by creating a transformation rule set using a list created by removing the duplicates from correctlabels.

```
relabelrules2 = Thread[Union[correctlabels] ->
                       Range[Length[Union[correctlabels]]]]
```

The correctly labeled matrix is therefore determined by repeatedly applying relabelrules1 to ul and then applying relabelrules2 to the result.

```
ufinal = ul //. relabelrules1 /. relabelrules2
```

THE CLUSTER LABELING PROGRAM

```
In[1]:= ClusterLabel[r_List] :=
        Module[{uup := u[[q-1, k]], uback := u[[q, k-1]],
              ulup := ul[[q-1, k]], ulback := ul[[q, k-1]]},
            AddZeros[w_] :=
                Prepend[w, Table[0, {Length[w[[1]]]}]];
            u = Transpose[AddZeros[Transpose[AddZeros[r]]]];
            ul = u /. 1 -> 0;
            ulp = {};
            Do[
              Which[u[[q, k]] == 1 && uup == 1 &&
                      uback == 1 && ulup =!= ulback,
                      ul[[q, k]] =
                        Min[ulp[[ulback]], ulp[[ulup]]];
                      ulp[[Max[ulp[[ulback]],ulp[[ulup]]]]] =
                              Min[ulp[[ulback]], ulp[[ulup]]],
                    u[[q, k]] == 1 && uup == 1,
                      ul[[q, k]] = ulup,
                    u[[q, k]] == 1 && uback == 1,
                      ul[[q, k]] = ulback,
                    u[[q, k]] == 1,
                      ul[[q, k]] = Max[ul] + 1;
                      AppendTo[ulp, Max[ul]]
                    ],
              {q, 2, Length[u]}, {k, 2, Length[u]}
              ];

        new = Range[Length[ulp]];
        relabelrules1 = Thread[new -> ulp];
        correctlabels = ulp//.relabelrules1;
        relabelrules2 = Thread[Union[correctlabels] ->
                Range[Length[Union[correctlabels]]]];
        ufinal = ul //. relabelrules1 /. relabelrules2
        ]
```

5.3 ■ RUNNING AND VISUALIZING THE PERCOLATION PROGRAM

We can use the ClusterLabel and SitePercolation programs to create and label a simple random site percolation system.

```
In[1]:= r = SitePercolation[0.493, 9];
        (cl = ClusterLabel[r]) //TableForm
```

```
Out[1]//TableForm= 0   0   0   0   0   0   0   0   0   0
                   0   1   0   2   0   0   0   0   3   3
                   0   1   0   2   0   0   4   0   3   3
                   0   0   0   0   0   5   0   0   3   3
                   0   0   0   0   5   5   0   3   3   3
                   0   0   6   6   0   0   3   3   0   0
                   0   7   0   6   0   6   0   3   0   0
                   0   7   0   6   6   6   6   0   0   0
                   0   0   8   0   0   6   6   0   0   9
                   0   0   8   8   0   0   0   0   9   9
```

The percolation cluster system can be represented using a `DensityPlot` graphic. In this case, sites are colored according to their values; so if the color function is `GrayLevel`, then sites with a value of 0 are colored black (`GrayLevel[0]`) and sites with higher values are colored progressively lighter. Here is a `ListDensityPlot` created with the `c1` lattice above. (Due to the way in which graphics functions such as `ListDensityPlot` and `RasterArray` choose the origin in their display, it is necessary to use `Reverse[c1]` if we want to create a graphic with the same orientation as that given by the list `c1` above.)

```
In[2]:= ListDensityPlot[Reverse[c1],
            ColorFunction -> GrayLevel,
            Frame -> False]
```

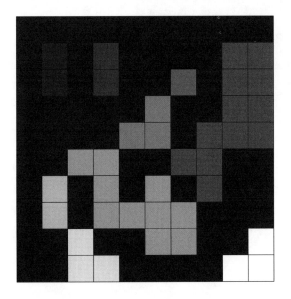

```
Out[2]= -DensityGraphics-
```

When it is desirable to use nonstandard colors, it might be preferable to display the resulting graphics using `RasterArray`. This has the added advantage of rendering much more quickly than graphics created with `ListDensityPlot`. This can be done with the following program:

```
In[3]:= ShowPercolation[cluster_List, opts___] := Module[{},
            Show[Graphics[RasterArray[
                    Reverse[Map[Hue[# Random[]/Max[cluster]]&,
                            cluster, {2}]]]],
                opts, AspectRatio -> 1]]
```

We can redisplay the c1 lattice using this program.

```
In[4]:= ShowPercolation[Reverse[c1]]
```

```
Out[4]= -Graphics-
```

Below, you can see what happens as the probability p is increased from 0.2 to 0.6 to 0.8. As p is increased, the average size of the clusters (shown in black here) increases.

In[5]:= **r = SitePercolation[0.2, 30];**
        ShowPercolation[ClusterLabel[r]]

Out[5]= –Graphics–

In[6]:= **r = SitePercolation[0.6, 30];**
        ShowPercolation[ClusterLabel[r]]

Out[6]= –Graphics–

```
In[7]:= r = SitePercolation[0.8, 30];
        ShowPercolation[ ClusterLabel[r] ]
```

```
Out[7]= -Graphics-
```

5.4 ■ COMPUTER SIMULATION PROJECTS

1. Consider a random site percolation system on a grid of size 50 × 50.

A. Determine the threshold value p_c, of p, at which a cluster spans the lattice.

B. The probability that an occupied site belongs to a spanning cluster is given by the number of sites in the spanning cluster divided by the total number of occupied sites. Determine the value for which $p = p_c$.

C. The *mean cluster size* is defined as the sum of $(s^2 n_s)$ divided by the sum of $(s n_s)$ where the summations are taken over all values of s, and n_s is the number of clusters having s sites. Determine the mean cluster size of the system.

2. In general, it is not a good idea to use an interactive computer simulation program since time is lost while the program asks for input from the user and waits for a response. However, this type of program can be extremely useful in the early stages of exploring a model when you have no good idea as to the proper range of parameter values that are needed.

Below is a program which uses interactive menu calls to draw continuum percolation site configurations and permit the visual determination of the percolation threshold.

The basic idea is to take a cookie sheet and randomly place cookies on the sheet until there is an uninterrupted path of cookies from one side of the sheet to the opposite side. The steps in the algorithm for doing this are:

i. Create a square with sides of length *s*.
ii. Place *n* disks of radius *r* on the square at random locations.
iii. Show the graphics of the disks on the square.
iv. Have the program ask, "How many more disks should be added to the square?" Note: At this point, the graphics should be examined by the user to see if there is an uninterrupted path of overlapping discs from one side of the square to the opposite side. If there is such a path, the appropriate response is 0. If there is no such path, the appropriate response is a non-zero integer.
v. If the response is a non-zero integer, place that number of additional disks randomly on the square and repeat the sequence of steps iii through iv. If the response is zero, terminate the program.

The continuum percolation site program is written as:

```
SwissCheese[s_Integer, n_Integer, r_] :=
   Module[{adddisks = 1},
      diskcount = n;
      t = Random[Real, {.15, .5}];
      box = Graphics[{RGBColor[.76, .61, .07],
                      Rectangle[{0, 0}, {s, s}]}];
      disklocs = Table[{Random[Real, s, 4],
                      Random[Real, s, 4]}, {n}];
      disks := ListPlot[disklocs,
               PlotStyle -> {RGBColor[0.65, 0.08, 0.7],
               PointSize[t/s]},
               Axes -> None,
               DisplayFunction -> Identity];
      While[adddisks != 0,
         Show[box, disks,
            DisplayFunction->$DisplayFunction];
         Print["total number of disks = ", diskcount];
         Print["disk density = ",
            (N[Pi])(r^2)(diskcount)/(s^2)];
         Print["PointSize = ", t];
            adddisks = Input["How many more disks?"];
         diskcount += adddisks;
         disklocs = Join[disklocs,
                     Table[{Random[Real, s, 4],
                         Random[Real, s, 4]},
                      {adddisks}]]
      ]
   ]
```

Using the SwissCheese program, determine the average number of disks needed to make an uninterrupted path of disks from one side of the sheet to the opposite side, as a function of the disk size.

3. A simulation has recently been proposed (see *Science News*, April 2, 1994, 218–219, and *Nature*, March 3, 1994, 368, 22) to model the formation of nanostructures by depositing on a surface particles which diffuse and aggregate. The effects of deposition, diffusion, and aggregation are distinguished by employing three steps in the program:

(1) Deposition: Particles are initially placed randomly on an $s \times s$ square lattice.

(2) Diffusion: At each time step, a randomly selected cluster of connected particles is moved one unit up, down, left, or right, with probability p.

(3) Aggregation: If particles end up adjacent to one another, their clusters stick together.

Write a program to model this process and create an animation showing that perfectly random motion can lead to self-organization.

REFERENCES

* P. G. de Gennes. Percolation—a new unifying concept. *La Recherche* 7 (1980) 919–26.

P. G. de Gennes. Percolation: quelques systemes nouveaux. *J. de Physique Colloque* 41 (1980) 17–26.

G. R. Grimmet. *Percolation.* Springer-Verlag New York 1989.

H. Kesten. *Percolation Theory for Mathematicians.* Birkhauser Boston 1982.

D. Stauffer and A. Aharony. *Introduction to Percolation Theory*, Second Edition. Taylor and Francis 1992.

CHAPTER 6

The Ising Model

INTRODUCTION

The Ising system is a model of imitative behavior in which individuals (*e.g.*, atoms or animals) modify their behavior so as to conform to the behavior of other individuals in their vicinity. In physics, the Ising model has been used to model ferromagnetism, phase separation in binary alloys, and spin glasses. In biology, it can model neural networks, flocking birds, fish schools, flashing fireflies, beating heart cells, and spreading diseases. The Ising model might also apply to fashion fads. We will present a probabilistic version of the Ising model, employing the Metropolis method (a deterministic version of the model will be described in Chapter 10).

6.1 THE PROBABILISTIC ISING MODEL

The probabilistic Ising model employs a constant temperature condition known as the *canonical ensemble formulation*. The two-dimensional version of the model consists of an $n \times n$ square lattice in which each lattice site has associated with it a value of 1 (*up spin*) or -1 (*down spin*). Spins on adjacent, nearest-neighbor lattice sites (henceforth called neighbors) interact in a pair-wise manner with a strength J known as the *exchange energy* or *constant*. When J is positive, the energy is lower when spins are in the same direction, and when J is negative, the energy is lower when spins are in opposite directions. There may also be an external field of strength B, known as the *magnetic field*. The *magnetization* of the system is the difference between the number of up and down spins in the lattice. In the spin-flipping process, a lattice site is randomly selected and it is either flipped (the sign of its value is changed) or not, based on the energy change in the system that would result from the flip.

THE ISING ALGORITHM

The sequence of steps **2** through **4** will be executed a number of times, first using the initial lattice configuration, and then using the lattice configuration resulting from the previous run-through of the sequence. We will describe the steps in terms of an arbitrary lattice configuration, called `lat`.

1. Create an $n \times n$ square lattice of randomly chosen site values of +1 and −1.
2. Select a random lattice site in `lat`.
3. Determine the energy change involved in flipping the spin at the selected lattice site. This is done in a number of steps:

   A. The neighbors to the selected site are determined. When the selected site is in the interior of `lat`, the neighbors are the sites north (above), south (below), west (left), and east (right) of the site. When the selected site is along the border of `lat`, some neighbors are taken from the opposing side of the lattice.
   Note: This way of choosing neighbors for border sites is known as the *periodic boundary condition.*
   B. The energy change that would result from flipping the spin of the selected lattice site is determined by the quantity 2 × (value of selected site) × ($B + J$ × (total spin from neighbors)), where B and J are input values given in $(1/kT)$ units, where k is the Boltzmann constant and T is the temperature.

4. Decide whether to flip the spin of the selected lattice site using the Metropolis method as follows:

   A. Check whether there is a negative energy change as a result of the flip.
   B. If the energy change is non-negative, check whether the exponential of (−energy change) is greater than a random number between 0 and 1.
   C. If one of these conditions is satisfied, flip the spin.
   D. Return the new `lat` value.

5. Execute the sequence of steps **2** through **4** m times.
6. Create a sublist, `monteCarloStepLis`, containing every (n^2)th element from the list of `lat` configurations.
   Note: The use of every (n^2)th element corresponds to giving each lattice site an equal chance to be selected. Each element in `monteCarloStepLis` is said to correspond to one monte carlo step.
7. For each element in `monteCarloStepLis`, calculate some global property of the lattice, such as the the absolute value of the magnetization of the lattice.

IMPLEMENTATION OF THE ISING ALGORITHM

1. First, an initial $n \times n$ lattice of spins is created, where p is the probability of an down spin and $1 - p$ is the probability of a up spin.

```
Table[2 Floor[p+Random[]]-1, {n},{n}]
```

2. A lattice site is randomly selected using

```
{i1, i2} = {Random[Integer,{1,n}], Random[Integer,{1,n}]}
```

3. Calculate the energy change involved in flipping the spin at the selected lattice site in a number of steps:

A. The neighbors to the selected site are `lat[[dn, i2]]`, `lat[[up, i2]]`, `lat[[i1, rt]]`, and `lat[[i1, lt]]`, where `lat[[dn, i2]]` is the lattice site beneath the site, `lat[[up, i2]]` is the lattice site above the site, `lat[[i1, rt]]` is the lattice site to the right of the site, and `lat[[i1, lt]]` is the lattice site to the left of the site. The values of `dn`, `up`, `rt`, and `lt` are determined as follows:

```
If[i1 == n, dn = 1, dn = i1 + 1];
If[i2 == n, rt = 1, rt = i2 + 1];
If[i1 == 1, up = n, up = i1 - 1];
If[i2 == 1, lt = n, lt = i2 - 1]
```

B. The energy change resulting from flipping the spin of the selected lattice site is given by:

```
2 lat[[i1, i2]] (B + J nnvalsum)
```

where `nnvalsum` is the total spin from the neighbors:

```
nnvalsum = lat[[dn, i2]] + lat[[up, i2]] +
                lat[[i1, rt]] + lat[[i1, lt]]
```

The value of `lat[[i1, i2]]` is either 1 or -1 and the value of `nnvalsum` is either 4, 2, 0, -2, or -4 (4 corresponds to all spins up, 2 to three spins up and one spin down, 0 to two spins up and two spins down, -2 to three spins down and one spin up and -4 to all spins down). Thus, there are ten possible energy changes as a result of spin flipping.

Rather than spending time calculating the energy change for flipping the selected site, we create a look-up table of the possible energy changes, where the arguments of `energydiff` are the value of a site and the sum of the values of its neighbors.

```
energydiff[ 1, 4] =  2 (B + 4J);
energydiff[ 1, 2] =  2 (B + 2J);
energydiff[ 1, 0] =  2 (B + 0J);
energydiff[ 1,-2] =  2 (B - 2J);
```

```
energydiff[ 1,-4] =  2 (B - 4J);
energydiff[-1, 4] = -2 (B + 4J);
energydiff[-1, 2] = -2 (B + 2J);
energydiff[-1, 0] = -2 (B + 0J);
energydiff[-1,-2] = -2 (B - 2J);
energydiff[-1,-4] = -2 (B - 4J);
```

By convention, a negative `energydiff` corresponds to a decrease of the lattice energy and a positive `energydiff` corresponds to an increase of the lattice energy.

4. Decide whether to flip the spin of the selected lattice site using the following criteria:

A. The energy of the lattice is lowered as a result of the flip.

```
energydiff[lat[[i1, i2]], nnvalsum] < 0
```

B. The energy of the lattice is raised as a result of the flip, but the exponential of (-`energydiff`) is greater than a random number between 0 and 1.

```
Random[] < Exp[-energydiff[lat[[i1, i2]], nnvalsum]
```

Note: As the energy is raised by the flip, the positive value of `energydiff` will increase, the exponential of (-`energydiff`) will decrease, becoming less likely to be greater than `Random[]`.

C. If either of these conditions are met, the spin of the selected site is flipped.

```
lat[[i1, i2]] = -lat[[i1, i2]]
```

D. The final lattice configuration is returned as `lat`.

The Metropolis method given in steps **4.A–D** is expressed using a conditional function.

```
If[energydiff[lat[[i1, i2]], nnvalsum] < 0 ||
    Random[] < Exp[-energydiff[lat[[i1, i2]], nnvalsum],
  lat[[i1, i2]] = -lat[[i1, i2]]; lat,
  lat]
```

We can combine steps **2** through **4** in an anonymous function, using the symbol # to represent the lattice configuration.

```
flip =
(lat = #;
 {i1, i2} = {Random[Integer,{1,n}],Random[Integer,{1,n}]};
          If[i1 == n, dn = 1, dn = i1 + 1];
          If[i2 == n, rt = 1, rt = i2 + 1];
          If[i1 == 1, up = n, up = i1 - 1];
          If[i2 == 1, lt = n, lt = i2 - 1];
          nnvalsum = lat[[dn, i2]] + lat[[up,i2]] +
                          lat[[i1, rt]] + lat[[i1, lt]];
          If[energydiff[lat[[i1, i2]], nnvalsum] < 0 ||
      Random[] < Exp[-energydiff[lat[[i1, i2]], nnvalsum]],
              lat[[i1, i2]] = -lat[[i1, i2]]; lat, lat]
)&
```

5. The repeated application of the sequence of steps **2** through **4**, *m* times, building a list of the lattice configuration after each flip attempt, is performed using the NestList function with the starting lattice configuration and the flip function.

```
flipLis = NestList[flip, lat, m]
```

6. A sublist monteCarloStepLis, containing every (n^2)th element from the list of lattice configurations is created.

```
monteCarloStepLis =  flipLis[[Range[1, m, n^2]]]
```

We can construct the program for the Ising model from these code fragments.

THE ISING PROGRAM

```
In[1]:= IsingMetropolis[n_, m_, B_, J_, p_] :=
        Module[{energydiff,initconfig,flip,flipLis},

          energydiff[ 1, 4] =  2 (B + 4J);
          energydiff[ 1, 2] =  2 (B + 2J);
          energydiff[ 1, 0] =  2 (B + 0J);
          energydiff[ 1,-2] =  2 (B - 2J);
          energydiff[ 1,-4] =  2 (B - 4J);
          energydiff[-1, 4] = -2 (B + 4J);
          energydiff[-1, 2] = -2 (B + 2J);
          energydiff[-1, 0] = -2 (B + 0J);
          energydiff[-1,-2] = -2 (B - 2J);
          energydiff[-1,-4] = -2 (B - 4J);

          initconfig = Table[2 Floor[p+Random[]]-1, {n},{n}];
```

```
flip = (
        lat = #;
        {i1, i2} =
          {Random[Integer,{1, n}], Random[Integer,{1, n}]};
        If[i1 == n, dn = 1, dn = i1+1];
        If[i2 == n, rt = 1, rt = i2+1];
        If[i1 == 1, up = n, up = i1-1];
        If[i2 == 1, lt = n, lt = i2-1];
        nnvalsum = lat[[dn, i2]] + lat[[up, i2]] +
                        lat[[i1, rt]] + lat[[i1, lt]];
        If[energydiff[lat[[i1, i2]],nnvalsum] < 0 ||
      Random[] < Exp[-energydiff[lat[[i1, i2]], nnvalsum]],
          lat[[i1, i2]] = -lat[[i1, i2]]; lat, lat]
        )&;

flipLis = NestList[flip, initconfig, m];
flipLis[[Range[1, m, n^2]]
]]
```

6.2 ■ MAGNETIZATION BEHAVIOR OF THE ISING MODEL

The magnetization behavior over time can be computed and then displayed using
the following program.

```
In[2]:= ShowIsingMagnetization[list_] :=
        Module[{n = Length[list[[1]]],
               longRangeOrderList},
           longRangeOrderList :=
             Map[Abs[Apply[Plus,Flatten[#]]/n^2]&, list];
           ListPlot[longRangeOrderList,
             PlotJoined -> True,
             PlotRange -> {0, 0.6},
             AxesLabel -> {FontForm["step", {"Times-Italic",8}],
                          FontForm["M", {"Times-Italic",8}]},
             PlotLabel -> FontForm["Long-range order parameter",
                                   {"Helvetica",10}]]]
```

6.3 ■ RUNNING THE MAGNETIZATION PROGRAM

Performing the flip operation 250,000 times on a 50×50 lattice system for $B/kT = 0$,
$J/kT = -0.1$, and $p = 0.2$, the relaxation of the magnetization is shown below. (To
be more precise, the graph actually displays the variation of the magnetization from 1
to 100 Monte Carlo steps, where one Monte Carlo step is equivalent to 2500 real-time
steps.)

```
In[3]:= ShowIsingMagnetization[
            IsingMetropolis[50, 250000, 0, -0.1, 0.2]]
```

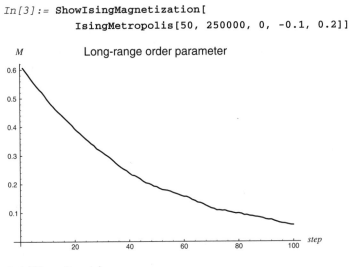

```
Out[3]= -Graphics-
```

6.4 ■ COMPUTER SIMULATION PROJECTS

1. **ListDensityPlot** can be used to see the state of the Ising model at any time step, using the following program.

```
ShowIsing[list_, opts___] :=
    ListDensityPlot[list,
                    opts, ColorFunction :>
                        (If[# == 0, Hue[.72], Hue[.28]]&)]
```

Using this function, create an animation showing the spin-flipping process.

2. Look at the size of clusters of spins having the same direction in the Ising model using the cluster labeling program developed in Chapter 5.

3. One variant of the Ising model (which has been used to model a lattice gas) replaces the Ising spin flip dynamics with spin exchange dynamics in which a pair of nearest neighbor sites is selected, the energy change resulting from interchanging their spins is determined, and the interchange decision is made using the Metropolis algorithm. (Note that the use of spin exchange dynamics conserves the number of spins in the lattice.) Implement spin exchange dynamics within the Ising model.

REFERENCES

* Earl Callen and Don Shapero. A theory of social imitation. *Physics Today* 27 (1974) 23–28.

Stephen G. Brush. History of the Lenz-Ising model. *Rev. Mod. Phys.* 39 (1967) 883–893.

CHAPTER 7
Darwinian Evolution

INTRODUCTION

Our understanding of biological evolution comes from the research of Charles Darwin whose work stands as one of the greatest achievements of scientific thinking. While the essence of the Darwinian theory—the idea of natural selection—remains unchallenged today, some of the details have been fine-tuned to better fit the historical record of evolution. Specifically, fossil findings indicate that, rather than evolving gradually over time, species evolve in an episodic manner, with intermittent bursts of evolutionary activity separating long periods of stasis in which little evolutionary change occurs.

While "punctuated equilibrium" is an experimental fact, its cause has been unknown. Recently, however, an extremely simple computer model of evolution has shown that the occurrence of "co-evolution," where the evolution of a species affects the evolution of the species with which it interacts, results in punctuated equilibrium behavior.

7.1 ■ THE CO-EVOLUTION MODEL

A one-dimensional lattice of size n, with periodic boundary conditions, is used to represent the ecosystem. Each lattice site represents a species and has a value given by a random number; the nearest neighbors of a site represent other species with which the species interacts. The random number value is a measure of the *fitness* of a species. This fitness represents the evolutionary barrier that must be overcome for the species to evolve. (This barrier is related to the amount of genetic material that must be modified for a species to mutate.) The larger the barrier, the more stable the species, and evolution proceeds by the mutation and natural selection of the least-fit species. In a given time step, the evolution of the least-fit species to a new fitness level also affects the fitness of the species with which it interacts (*e.g.*, by food chains, predator-prey, and parasite-host relationships). The evolutionary process is represented by finding the site with the lowest random number and generating new random numbers for all three sites in the neighborhood of that site.

THE CO-EVOLUTION ALGORITHM

1. Create a one-dimensional lattice of size n, in which each lattice site has a value given by a random number. Also create an empty list, called leastFitSites.

2. Determine the location of the site in the lattice with the lowest value.

3. Place the location of the site with the lowest value into the leastFitSites list and replace the values of the lowest-valued site and its nearest-neighbor sites on either side with new random numbers.

4. Repeat steps **2** through **3** a specified number of times.

IMPLEMENTATION OF THE CO-EVOLUTION ALGORITHM

The sequence of steps **2** through **3** will be executed a number of times, first using the value of the ecosystem given by prebiotic in step **1**, and then using the value of the ecosystem resulting from the previous run-through of the sequence. We'll describe the sequence in terms of an arbitrary ecosystem configuration, which we'll call eco.

1. The initial ecosystem of n species is first constructed.

```
prebiotic = Table[Random[], {n}]
```

The empty list leastFitSites is then created.

```
leastFitSites = {}
```

2. The species with the lowest fitness value is determined using the following:

```
Position[eco, Min[eco]]
```

3. The neighborhood of the least-fit site is determined by applying the anonymous function

```
Join[# - 1 /. 0 -> n,
     #,
     # + 1 /. (n + 1) -> 1]&
```

to the position of the least-fit site:

```
Join[# - 1 /. 0 -> n,
     #,
     # + 1 /.(n + 1) -> 1]&[Position[eco, Min[eco]]]
```

Note: The transformation rules are used to implement the periodic boundary conditions so that when the least-fit site is the leftmost (rightmost) site on the lattice, the nearest neighbor site on the left (right) is taken from the right (left) end of the lattice.

The center site location is placed in the `leastFitSites` list and the values of the three neighborhood sites are replaced by random numbers.

```
ReleaseHold[
  ReplacePart[eco,
              Hold[Random[]],
              Join[# - 1 /. 0 -> n,
                  AppendTo[leastFitSites, #];
                  #,
                  # + 1 /. (n+1) -> 1]&[Position[eco, Min[eco]]]
              ]]
```

Note: The `Hold` function is wrapped around `Random[]` to prevent it from being evaluated before it is substituted into the `eco` list (otherwise the same random number value would be placed in the three locations in the list). The resulting list looks like:

```
{..., Hold[Random[]], Hold[Random[]], Hold[Random[]], ...}
```

Applying `ReleaseHold` to the above list results in a different random number being generated for each of the three sites.

We can combine steps **2** through **3** in an anonymous function, using the symbol y to represent the ecosystem configuration.

```
In[1]:= fitness =
        Function[y,
            ReleaseHold[ ReplacePart[y, Hold[Random[]],
                Join[# - 1 /. 0 -> n,
                    AppendTo[leastFitSites, #]; #,
                    # + 1 /. (n+1) -> 1]&[Position[y, Min[y]]]
            ]]]
```

4. The anonymous function `fitness`, is repeatedly applied *t* times to the ecosystem using the `Nest` function.

```
Nest[fitness, prebiotic, t]
```

The program for Darwinian evolution is created from these code fragments.

THE CO-EVOLUTION PROGRAM

```
In[1]:= DarwinianEvolution[n_, t_] :=
          Module[{prebiotic, fitness, leastFitSites},
             leastFitSites = {};
             prebiotic = Table[Random[], {n}];
             fitness = Function[y, ReleaseHold[
                ReplacePart[y, Hold[Random[]],
                   Join[# - 1 /. 0 -> n,
                      AppendTo[leastFitSites, #];
                      #,
                      # + 1 /. (n+1) -> 1]&[Position[y, Min[y]]]
                ]]];
             Nest[fitness, prebiotic, t];
             Flatten[leastFitSites]
          ]
```

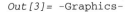 RUNNING THE PROGRAM

We will look at the last 6000 generations of a 200-species ecosystem evolving over 8000 generations.

Looking at a plot of mutation site location versus generation, we can clearly see nonrandom behavior.

```
In[2]:= mutationSites = Drop[DarwinianEvolution[200, 8000], 2000];
```

```
In[3]:= ListPlot[mutationSites]
```

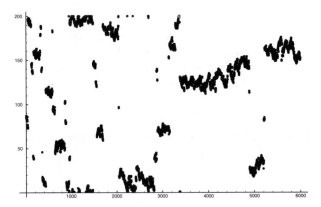

Out[3]= -Graphics-

Widely separated clusters of nearby (*i.e.*, interacting) sites change their values in an intermittent fashion. This is distinctly different from the behavior of an equal-sized sample of random numbers, which would produce a more uniform plot of filled regions.

We can calculate the frequency with which the values of sites change using the following function.

In[4]:= **Frequency[lis_] := Map[{#, Count[lis, #]}&, Union[lis]]**

A graphic of the result again shows nonrandom behavior.

In[5]:= **ListPlot[Frequency[mutationSites]]**

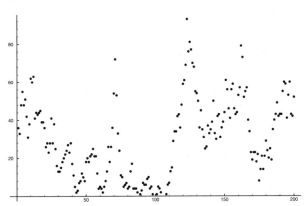

Out[5]= -Graphics-

Some sites do not change at all while others change many times. This differs from the behavior of an equal-sized sample of random numbers, which would produce a uniform, narrow band of points (6000 random numbers between 1 and 200 form a band of points centered around the value 30).

According to the co-evolution model, interacting (co-evolving) species tend to evolve in an intermittent (punctuated equilibrium) fashion. Moreover, species that have evolved recently (*Homo sapiens*, for example) are more likely to evolve again than species that have remained unchanged for long periods (sharks and cockroaches). Generally, the model's results indicate that the oft-heard expression "survival of the fittest" might be restated as "evolution of the least fit."

7.3 ◼ COMPUTER SIMULATION PROJECTS

1. One quantity of interest is distribution of the distances between successive mutation sites. The distances between successive mutations can be computed using the following anonymous function:

```
ConsecutiveMutationsDist = Function[{y, z},
    Drop[Abs[z Floor[2#/z] - #]&[Abs[y -
                RotateRight[y]]], 1]]
```

Here `y` is the list of mutation sites and `z` is the total number of species in the ecosystem. The periodic boundary conditions of the model turns the line into a circle, so that the distance between a pair of points along the line can be measured in two directions. The anonymous `ConsecutiveMutationsDist` function determines the shortest of these distances.

Using the `ConsecutiveMutationsDist` function and the `frequency` function, compute the power law dependence of the distribution of distances between successive mutations.

Hints: The Darwinian evolution model shows noninteger power law behavior. This is indicative of self-organized critical (SOC) behavior. According to this view, ecological catastrophes such as the mass extinction of the dinosaurs might simply be the result of intrinsic factors characteristic of the SOC, rather than due to extrinsic factors such as meteorite impact or volcanic eruption. (This subject is further discussed in Chapter 9.)

REFERENCES

* Per Bak, Henrik Flyvberg, and Kim Sneppen. Can we model Darwin? *New Scientist* 12 (March 1994) 36–39.

* Towards a statistical mechanics of biological evolution? *Physics World* (March 1994) 24–25.

* John Maddox. Punctuated equilibrium by computer. *Nature* 371 (1994) 197.

Per Bak and Kim Sneppen. Punctuated equilibrium and criticality in a simple model of evolution. *Phys. Rev. Letters* 71 (1993) 4083–4086.

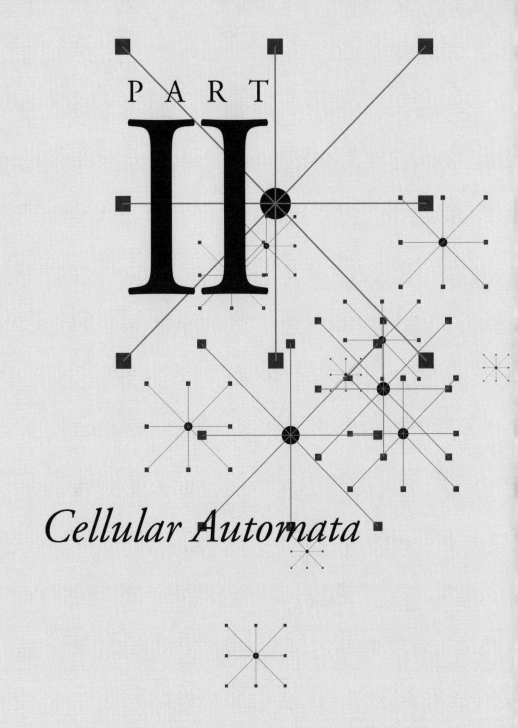

II

Cellular Automata

Cellular Automata Preliminaries

What is a Cellular Automaton?

A *cellular automaton* (or CA) consists of a discrete system of lattice sites having various initial values. These sites evolve in discrete time steps as each site assumes a new value based on the values of some local neighborhood of sites and a finite number of previous time steps.

Before starting to work with cellular automata models, we need to provide some background. We will define the lattices that we'll be using and show the neighborhoods of lattice sites for various boundary conditions.

CA Lattices

In one dimension, a cellular automaton is a simple linear list of the following form (where expr evaluates to numbers and/or symbols):

```
Table[expr, {i, 1, s}]
```

In two dimensions, various lattices can be used (*e.g.*, rectangular, triangular, hexagonal). We will be using rectangular $n \times m$ lattices that can be written using the Table function.

```
Table[expr, {i, 1, n}, {j, 1, m}]
```

CA NEIGHBORHOODS

The *neighborhood* of a lattice site consists of the site and its nearest neighbor sites (called *neighbors*). Two kinds of neighborhoods are commonly defined for the rectangular lattice.

- A **von Neumann** neighborhood consists of the site and the four nearest neighbors, north (above), east (right), south (below), and west (left) of the site. In the graphics below, the site is represented by a filled disk and its neighbors by gray squares.

von Neumann neighborhood

- A **Moore** neighborhood consists of the site and the eight nearest neighbor sites, north, northeast, east, southeast, south, southwest, west, and northwest of the site.

Moore neighborhood

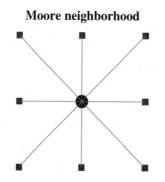

The nearest neighbors of sites along the sides (borders) of a lattice are determined differently for various boundary conditions. To illustrate these different criteria, the corresponding Moore neighborhoods are shown below for each site in the following simple lattice:

1	2	3
4	5	6
7	8	9
10	11	12

PERIODIC BOUNDARIES

- The nearest neighbor left of a site on the left border is the site in the same row on the right border.
- The nearest neighbor right of a site on the right border is the site in the same row on the left border.
- The nearest neighbor above a site on the top border is the site in the same column on the bottom border.
- The nearest neighbor below a site on the bottom border is the site in the same column on the top border.

```
12 10 11    10 11 12    11 12 10
 3  1  2     1  2  3     2  3  1
 6  4  5     4  5  6     5  6  4

 3  1  2     1  2  3     2  3  1
 6  4  5     4  5  6     5  6  4
 9  7  8     7  8  9     8  9  7

 6  4  5     4  5  6     5  6  4
 9  7  8     7  8  9     8  9  7
12 10 11    10 11 12    11 12 10

 9  7  8     7  8  9     8  9  7
12 10 11    10 11 12    11 12 10
 3  1  2     1  2  3     2  3  1
```

As the table above shows, the neighbors of site 4 are 3, 1, 2, 6, 5, 9, 7, and 8.

ABSORBING BOUNDARIES

- The nearest neighbor left of a site on the left border is zero.
- The nearest neighbor right of a site on the right border is zero.
- The nearest neighbor above a site on the top border is zero.
- The nearest neighbor below a site on the bottom border is zero.

```
0   0   0      0   0   0      0   0   0
0   1   2      1   2   3      2   3   0
0   4   5      4   5   6      5   6   0

0   1   2      1   2   3      2   3   0
0   4   5      4   5   6      5   6   0
0   7   8      7   8   9      8   9   0

0   4   5      4   5   6      5   6   0
0   7   8      7   8   9      8   9   0
0  10  11     10  11  12     11  12   0

0   7   8      7   8   9      8   9   0
0  10  11     10  11  12     11  12   0
0   0   0      0   0   0      0   0   0
```

So, using absorbing boundaries, the nearest neighbors of the number 4 in the original lattice are 0, 1, 2, 0, 5, 0, 7, 8.

SKEWED BOUNDARIES

- The nearest neighbor left of the first lattice site is the last lattice site.
- The nearest neighbor left of any other site on the left border is the site on the right border in the preceding row.
- The nearest neighbor right of the last lattice site is the first site.

```
 9  10  11     10  11  12     11  12   1
12   1   2      1   2   3      2   3   4
 3   4   5      4   5   6      5   6   7

12   1   2      1   2   3      2   3   4
 3   4   5      4   5   6      5   6   7
 6   7   8      7   8   9      8   9  10

 3   4   5      4   5   6      5   6   7
 6   7   8      7   8   9      8   9  10
 9  10  11     10  11  12     11  12   1

 6   7   8      7   8   9      8   9  10
 9  10  11     10  11  12     11  12   1
12   1   2      1   2   3      2   3   4
```

- The nearest neighbor right of any other site on the right border is the site on the left border in the following row.
- The nearest neighbor above any site on the top border is the site in the same column in the last row.

- The nearest neighbor below any site on the bottom border is the site in the same column in the first row.

WRITING CELLULAR AUTOMATON RULES IN *MATHEMATICA*

Mathematica has two features that can be used to advantage in creating cellular automata rules: (1) Rules (or functions) can be defined with the `Listable` attribute, and (2) more specific rules are applied before more general rules.

The use of these features in writing CA rules is discussed in depth in Chapter 8 and the reader should begin their CA studies by reading that chapter before proceeding to the other CA chapters.

CHAPTER 8
The Game of Life

INTRODUCTION

The Game of Life, created by the British mathematician John Conway, is the most famous cellular automaton (CA). It has been said that more computer time has been spent on running the Game of Life than on any other computation. The Game of Life was the first program run on the Connection Machine, the world's first parallel computer. Most importantly, it is the forerunner of so-called *artificial life* (or *a-life*) systems which are of great interest today, not only for their biological implications, but for the development of so-called "intelligent agents" for computers.

8.1 ▪ THE GAME OF LIFE

The Game of Life is played on a two-dimensional square lattice with periodic boundary conditions. Lattice sites have a value of 0 or 1 (*i.e.*, the lattice is a Boolean matrix). A site with value 1 is said to be *alive* and a site with value 0 is said to be *dead*. The system evolves by updating all of the sites in the lattice simultaneously, based on their Moore neighborhoods, until two successive lattice configurations are identical, or until a specified number of updates (time steps) have occurred.

THE LIFE RULES

The Game of Life CA rules are based on the value of a site and the sum of the values of its neighbors. The rules, known as *life and death* rules, are:

- A living site with two living nearest neighbor sites remains alive.
- Any site with three living nearest neighbor sites stays alive or is born.
- All other sites either remain dead or die.

THE GAME OF LIFE ALGORITHM

1. The game is played on a square lattice, with the initial set-up given by:

```
initconfig = Table[Random[Integer], {s}, {s}]
```

2. The matrix whose elements represent the number of living nearest neighbors to the corresponding sites in the lattice mat, is given by:

```
In[1]:= livingNghbrs[mat_] := Apply[Plus,
            Map[RotateRight[mat, #]&,
               {{-1,-1}, {-1,0}, {-1,1}, {0,-1},
                {0,1}, {1,-1}, {1,0}, {1,1}}]]
```

The use of the livingNghbrs function can be demonstrated with a simple example.

```
In[2]:= (board = Table[Random[Integer],{4},{4}] //MatrixForm
```

```
Out[2]//MatrixForm= 0   1   1   1
                    0   1   0   1
                    0   0   0   0
                    1   0   1   0
```

```
In[3]:= (livingNghbrs[board]) //MatrixForm
```

```
Out[3]//MatrixForm= 5   4   5   4
                    4   2   5   2
                    3   3   3   3
                    2   4   3   4
```

Comparing the livingNghbrs[board] matrix with the matrix of the Moore neighborhoods of the initial configuration board, we see that each element in the livingNghbrs[board] matrix is the number of living neighbors to the corresponding site in the board matrix.

```
In[4]:= bc = Join[{Last[#]}, #, {First[#]}]&;
            Partition[bc[Map[bc, board]],
               {3, 3}, {1, 1}]//MatrixForm
```

```
Out[4]//MatrixForm= 0 1 0   1 0 1   0 1 0   1 0 1
                    1 0 1   0 1 1   1 1 1   1 1 0
                    1 0 1   0 1 0   1 0 1   0 1 0

                    1 0 1   0 1 1   1 1 1   1 1 0
                    1 0 1   0 1 0   1 0 1   0 1 0
                    0 0 0   0 0 0   0 0 0   0 0 0

                    1 0 1   0 1 0   1 0 1   0 1 0
                    0 0 0   0 0 0   0 0 0   0 0 0
                    0 1 0   1 0 1   0 1 0   1 0 1

                    0 0 0   0 0 0   0 0 0   0 0 0
                    0 1 0   1 0 1   0 1 0   1 0 1
                    1 0 1   0 1 1   1 1 1   1 1 0
```

3. In creating rewrite rules to express the life and death rules, it is important to be parsimonious, using as many rules as are needed, but no more. For example, we could write a rule for each possible neighborhood (*e.g.*, `update[{{0, 1, 0},` `{0, 1, 0}, {0, 0, 0}}] := 0`) but this approach would result in $2^9 = 512$ rules (since each of the nine sites can have two values) which would be far too many to enter or use. Instead, the life and death rules can be expressed in a minimal number of rules simply by translating them directly from words to code:

```
update[1, 2] := 1

update[_, 3] := 1

update[_, _] := 0

Attributes[update] := Listable
```

The first argument of `update` is the value of the center site of a neighborhood and the second argument is the number of living nearest neighbor sites.

The `update` function is given the `Listable` attribute so that when `update` is applied to a matrix of site values (*e.g.*, board) and a matrix of the number of living neighbors to these sites (*e.g.*, `livingNghbrs[board]`), it creates a matrix whose elements are the `update` function with the corresponding elements of the matrices as its arguments. This `Listable` behavior can be demonstrated using a general function, g, with the board and `livingNghbrs[board]` matrices.

```
In[5]:= Attributes[g] = Listable;
        g[board, livingNghbrs[board]] //MatrixForm
```

```
Out[5]//MatrixForm= g[0, 5]   g[1, 4]   g[1, 5]   g[1, 4]
                    g[0, 4]   g[1, 2]   g[0, 5]   g[1, 2]
                    g[0, 3]   g[0, 3]   g[0, 3]   g[0, 3]
                    g[1, 2]   g[0, 4]   g[1, 3]   g[0, 4]
```

Using the update rules with the board and livingNghbrs matrices, we see that each site in board has been correctly updated according to the update rules.

```
In[6]:= update[1, 2] := 1
        update[_, 3] := 1
        update[_, _] := 0
        Attributes[update] = Listable;
        update[board, livingNghbrs[board]] //MatrixForm
```

```
Out[6]//MatrixForm=  0    0    0    0
                     0    1    0    1
                     1    1    1    1
                     1    0    1    0
```

While the three update rules overlap with one another, there is no confusion about
when to use each rule, because *Mathematica* applies more specific rules before
more general rules. Thus, while a living site with two living nearest neighbor sites
will satisfy both the first and third rules, the first rule is used because it is the
most specific applicable rule. Similarly, while any site having three living nearest
neighbors will satisfy both the second and third rules, the second rule is used
because it is the most specific applicable rule. The third rule is more general than
the other two rules and hence is only used if neither of the other rules can be
used.

4. The lattice is updated until it no longer changes or until *t* time steps have been
taken. An anonymous function, update[#, livingNghbrs[#]]&, is used in
FixedPointList. Here, # represents the lattice configuration.

```
evolution = FixedPointList[update[#, livingNghbrs[#]]&,
                           initconfig, t]
```

Putting these code fragments together, we can write the Game of Life program.

THE GAME OF LIFE PROGRAM

```
In[1]:= LifeGame[s_, t_]:=
        Module[{initconfig, livingNghbrs, update},
            initconfig = Table[Random[Integer], {s}, {s}];
            livingNghbrs[mat_] :=
                Apply[Plus, Map[RotateRight[mat, #]&,
                            {{-1,-1},{-1,0},{-1,1},{0,-1},
                             {0,1},{1,-1},{1,0},{1,1}}]];
            update[1, 2] := 1;
            update[_, 3] := 1;
            update[_, _] := 0;
            Attributes[update] = Listable;
            FixedPointList[update[#, livingNghbrs[#]]&,
                           initconfig, t]
        ]
```

8.2 ■ RUNNING THE GAME OF LIFE PROGRAM

Here is the Game of Life, run for a small system—a 4 × 4 lattice run for three generations.

```
In[2]:= LifeGame[4, 3] //MatrixForm
```

```
Out[2]//MatrixForm=  0   0   1   0
                     1   1   0   1
                     0   0   0   1
                     0   1   1   0

                     0   0   0   0
                     1   1   0   1
                     0   0   0   1
                     0   1   1   1

                     0   0   0   0
                     1   0   1   1
                     0   0   0   0
                     1   0   1   1

                     0   0   0   0
                     0   0   0   1
                     0   0   0   0
                     0   0   0   1
```

When the Game of Life is run, we are usually interested in identifying various patterns of 1s amongst the 0s, and observing their behaviors. This is best done using a graphical, rather than numerical, display of the game board. A program for creating an animation of the Game of Life is given below.

```
In[3]:= ShowLife[list_, opts___Rule]:=
          Map[(Show[Graphics[RasterArray[
              Reverse[list[[#]]] /.
                  {1 -> RGBColor[1,0,0],
                   0 -> RGBColor[0,0,0]}]],
              AspectRatio->Automatic,
              opts]])&,
          Range[Length[list]]]
```

We can show all of the animation cells in a single display using `Graphics-Array`. The graphic on the next page shows 24 frames of the first 50 iterations by displaying every other frame. The game is played on a 50 × 50 lattice.

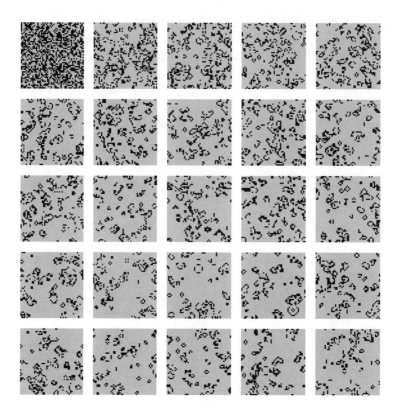

COMPUTER SIMULATION PROJECTS

1. The Game of Life is most interesting to watch when persistent patterns, known as *life-forms*, occur during the evolution process. One pattern that has been extensively studied is known as the *glider* and is defined as follows:

```
glider[x_, y_] := {{x, y}, {x+1, y}, {x+2, y},
                   {x+2, y+1}, {x+1, y+2}}
```

Another interesting life-form is the *beehive*.

```
beehive[x_, y_] := {{x, y}, {x, y+1}, {x, y+2}, {x, y+3},
                    {x, y+4}, {x, y+5}, {x, y+6}}
```

Modify the program for the Game of Life so that the lattice can be seeded with life forms.

2. Using the modified Life program from the first project above, observe the behavior over time of a glider and of a beehive.

Hints: The beehive life-form appears after a certain number of generations and remains unchanged thereafter, while the glider life-form appears, disappears and then reappears in a shifted position every few generations.

3. Rudy Rucker has suggested three CA models of various physical phenomena. Run the following CA programs and observe their behaviors.

A. Diffusion: The Melt CA models the spread of molecules in space (*e.g.*, perfume) or the spread of heat in a material.

The sites of a square lattice have values ranging from 0 to r (*e.g.*, $r = 255$), where r represents concentration or temperature. At each update, a site's value is replaced by the average of the eight nearest neighbors in its Moore neighborhood.

The `Melt` program is written as:

```
Melt[r_, s_, t_] := Module[{init, ngbrsAve},
    init = Table[Random[Integer, {0,r}], {s}, {s}];
    ngbrsAve[mat_] := Floor[Apply[Plus,
            Map[RotateRight[mat, #]&,
              {{-1,-1},{-1,0},{-1,1},{0,-1},
               {0,1},{1,-1},{1,0},{1,1}}]]/8];
    NestList[ngbrsAve, init, t]]
```

B. Boiling: The Rug CA is a variant of the Melt CA and models the transition from liquid to vapor. The sites of a square lattice have values ranging from 0 to $r - 1$ (*e.g.*, $r = 256$). At each update, a site's value is replaced by the average of its eight nearest neighbors plus one, unless the resulting value is r or higher, in which case, it goes to 0 (or some low value).

According to the rules, if at any time, every nearest neighbor of a site has the maximum value $r - 1$, the site's value becomes 0 in that time step. This zero-valued site will then lower the values of its nearest neighbors in the next time step. This process is analogous to the way in which a sufficiently hot region of water gives up some heat by forming a bubble of steam, which momentarily cools off the water around the bubble.

The `Rug` program is written as:

```
Rug[r_, s_, t_] := Module[{init, ngbrsAve},
    init = Table[Random[Integer, {0, r - 1}], {s}, {s}];
    ngbrsAve[mat_] := Floor[Apply[Plus,
            Map[RotateRight[mat, #]&,
              {{-1,-1},{-1,0},{-1,1},{0,-1},
               {0,1},{1,-1},{1,0},{1,1}}]]/8];
    NestList[Mod[ngbrsAve[#] + 1],r]&, init, t]]
```

C. Weathering: The Vote CA models the "smoothing-off" of jagged edges.

The sites of a square lattice have values of 0 or 1. At each update, a site's value is replaced by the value possessed by the majority of the nine sites in its Moore neighborhood, with the following exceptions: If there is one more neighborhood site with a value of 1 than with a value of 0 (*i.e.*, five sites with value 1 and four sites with value 0), the site value is updated to 0 and if there is one less neighborhood site with a value of 1 than with a value of 0 (*i.e.*, four sites with value 1 and five sites with value 0), the site value is updated to 1. The rules of this CA are given in the table below:

```
In[1]:= Print["                              The CA Vote Rule Table"];
        TableForm[{Range[0, 9],
            {0, 0, 0, 0, 1, 0, 1, 1, 1, 1}},
            TableHeadings ->
                {{"Sum over neighborhood", "New cell value"},
                None}]
```

The CA Vote Rule Table

Sum over neighborhood	0	1	2	3	4	5	6	7	8	9
New cell value	0	0	0	0	1	0	1	1	1	1

The program is written as:

```
VoteNearCallsToLosers[s_, t_]:=
    Module[{rule, init, ngbrhdTotal},
        init = Table[Random[Integer], {s}, {s}];
        ngbrhdTotal[mat_] :=
            Apply[Plus, Map[RotateRight[mat, #]&,
                    {{-1,-1},{-1,0},{-1,1},{0,-1},{0,1},
                    {0,0},{1,-1},{1,0},{1,1}}]];
        rule[5] := 0;
        rule[x_] := Floor[x/4];
        Attributes[rule] = Listable;
        NestList[rule[ngbrhdTotal[#]]&,init,t]]
```

REFERENCES

* William Poundstone. *The Recursive Universe*. Oxford University Press 1985.

Steven Levy. *Artificial Life: A Report from the Frontier Where Computers Meet Biology*. Random House 1993.

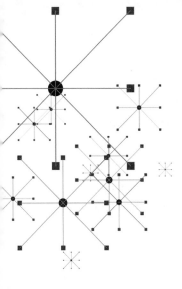

CHAPTER 9

Avalanches

INTRODUCTION

The behavior of many complex systems (large systems consisting of many components that have short-range interactions) appears to evolve naturally toward a meta-stable state, called the *self-organized critical* state. Once a system is in this state, even small changes in the system can cause chain reactions of all sizes and durations, affecting any number of the system's components and leading to a catastrophe. Self-organized criticality (SOC) behavior can account for changes that occur in many systems, including geology (earthquakes, volcanoes, and landscape formation), astronomy (pulsar glitches and solar flares), economy (stock market fluctuations) ecosystems (evolution) and fluid dynamics (turbulence). In this chapter we will develop the sandpile cellular automaton, which displays self-organized critical behavior.

9.1 ■ THE SANDPILE MODEL

The sandpile (or avalanche) model takes place on a two-dimensional square lattice with absorbing boundary conditions. Internal lattice sites have random values between 1 and 8 while border sites have a value of 0. Sites that have a value of 5 or more are said to be "top-heavy." The system evolves by simultaneously updating all of the sites based on their von Neumann neighborhoods, until there are no top-heavy sites left.

   Note: In the usual statement of the sandpile model, the initial values of lattice sites range from 0 and 7 and sites with a value of 4 or more are said to be "top heavy." We have changed this because we use 0 to handle the absorbing boundary conditions.

THE SANDPILE RULES

The sandpile CA rule is stated as follows:

- Any site with a threshold value of 5 or higher has its value reduced by 4, and the values of its four nearest neighbor sites are increased by 1.

When the rule is stated this way, in terms of how the value of a site affects both itself and its neighbors, it is not obvious how to update all of the sites simultaneously. The rule can be made more transparent by stating it entirely in terms of the effect that the values of a site and its neighbors have on the site:

- The value of a top-heavy site is increased by the number of nearest neighbor top-heavy sites minus four. The value of any other site is increased by the number of top-heavy nearest neighbor sites.

Looking at the rules, we can see that two extreme values occur in the "landscape": A site with a value of 4 and all top-heavy nearest neighbor sites will be updated to 8 while a site with a value of either 5 or 1 and no top-heavy nearest neighbor sites will be updated to 1.

THE SANDPILE ALGORITHM

1. A. The initial landscape configuration consists of an $(s + 2) \times (s + 2)$ lattice whose interior sites have randomly chosen integer values between 1 and 4 and whose border sites (sites in the first and last rows and first and last columns) have value 0.

```
landscape =
    absorbBC[Table[Random[Integer, {1, 4}], {s}, {s}]]
```

The function `absorbBC` is an anonymous function given as follows:

```
absorbBC =
    (Prepend[Append[Map[Prepend[Append[#, 0], 0]&, #],
                    Table[0, {Length[#] + 2}]],
            Table[0, {Length[#] + 2}]])&
```

B. Interior sites in `landscape` are randomly chosen and incremented by 1 until one of the sites has a value of 5.

```
While[Max[landscape]  <  5,
    randx = Random[Integer, {2, s + 1}, {2, s + 1}];
    randy = Random[Integer, {2, s + 1}, {2, s + 1}];
    landscape[[randx, randy]]++
    ]
```

2. A. The number of top-heavy neighbor sites (sites having a value greater than 4) is calculated using:

```
topHeavyNgbrs[mat_] :=
    Apply[Plus, Map[RotateRight[Floor[mat/5.0], #]&,
                    {{-1,0}, {0,-1}, {0,1}, {1,0}}]]
```

In this function, the quantity `Floor[mat/5.0]` changes the value of sites having a value less than 5 to 0 and the value of sites having value greater than or equal to 5 (*i.e.*, top-heavy sites) to 1. Hence, `topHeavyNgbrs[mat]` creates a matrix in which the value of a site is the number of top-heavy nearest neighbor sites to the corresponding site in `mat`.

B. The avalanching rules of the model are expressed in three statements:

```
update[r_, t_] := r + t /; r < 5

update[r_,  t_] := r - 4 + t

update[0, _] := 0

Attributes[update] = Listable
```

The first argument of `update` is the value of a site and the second argument of `update` is the number of top-heavy neighbors.

The first rule updates a site which is not top-heavy and has t top-heavy neighbors. The second rule updates a site which is top-heavy and has t top-heavy neighbors. The third rule maintains the absorbing boundary conditions.

3. The lattice is updated m times or until it stops changing (which occurs when there are no top-heavy sites) by repeatedly applying an anonymous function `update[#, topHeavyNgbrs[#]]&` to the landscape with `FixedPointList`:

```
FixedPointList[update[#, topHeavyNgbrs[#]]&, landscape, m]
```

Here, # represents the lattice configuration.

The program for the sandpile model is now easily constructed from its parts.

THE SANDPILE PROGRAM

```
In[1]:= Sandpile[s_, m_]:=
        Module[{absorbBC, landscape, topHeavyNgbrs, update},
            absorbBC =
              (Prepend[Append[Map[Prepend[Append[#, 0], 0]&, #],
                        Table[0, {Length[#] + 2}]],
                    Table[0, {Length[#] + 2}]])&;
            landscape = absorbBC[Table[Random[Integer, {1, 4}],
                                {s}, {s}]];
```

```
While[Max[landscape]  <  5,
     randx = Random[Integer, {2, s + 1}, {2, s + 1}];
     randy = Random[Integer, {2, s + 1}, {2, s + 1}];
     landscape[[randx, randy]]++
     ];

topHeavyNgbrs[mat_] :=
   Apply[Plus, Map[RotateRight[Floor[mat/5.0], #]&,
                   {{-1,0}, {0,-1}, {0,1}, {1,0}}]];

update[0, _] := 0;
update[r_, t_] := r + t /; r < 5;
update[r_, t_] := r - 4 + t;
Attributes[update] = Listable;

FixedPointList[update[#,topHeavyNgbrs[#]]&, landscape, m]
]
```

9.2 ■ RUNNING THE SANDPILE PROGRAM

The sandpile model is well-suited to a three-dimensional graphics display. We'll first load the Graphics3D.m package,

```
In[2]:= Needs["Graphics`Graphics3D`"]
```

Here is the animation program:

```
In[3]:= ShowCatastrophe[list_, opts___]:=
        Map[(BarChart3D[list[[#]],
                        opts, PlotRange -> {0,8}])&,
            Range[Length[list]]]
```

The PlotRange of the z-coordinate in the graphics is specified to run from the minimum to maximum values of a site.

A GraphicsArray resulting from running the program on a small system is shown below.

```
In[4]:= Show[GraphicsArray[Partition[
            ShowCatastrophe[Sandpile[5, 20],
                            Axes -> False,
                            Boxed -> False],
            3]]]
```

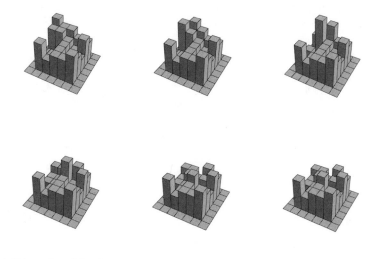

Out[4]= -GraphicsArray-

9.3 ■ Computer Simulation Projects

1. Two-dimensional graphics can also be used with the sandpile model. In this case, coloring can be used to distinguish between the various site values or heights. Calling the list resulting from running the sandpile program frames, we can use the ListDensityPlot function.

```
Map[(ListDensityPlot[frames[[#]]])&,
                     Range[Length[frame]]]
```

In this case, the shading is done for us; we can also use RasterArray and create our own coloring scheme:

```
colors = Thread[Rule[Range[0, 8],
                  Map[(RGBColor[0.1, 0.9,#])&,
              Table[N[1/Range[9]] ] ]]];

Map[(Show[Graphics[RasterArray[frames[[#]] /. colors],
          Axes -> None,
          AspectRatio -> 1]])&.
     Range[Length[frames]]]
```

Create and compare three-dimensional and two-dimensional graphical displays of the same sandpile program run.

REFERENCES

* Per Bak and Kan Chen. Self-organized criticality. *Scientific American* 264 (1991) 46–53.

Per Bak. Catastrophes and self-organized criticality. *Computers in Physics* 5 (July/Aug 1991) 430–433.

CHAPTER 10

The Q2R Ising Model

INTRODUCTION

The *Q2R* model is a CA version of the Ising model that was discussed in Chapter 6. The two models differ in several significant ways. Q2R uses a microcanonical ensemble (constant energy) condition, while the Metropolis Ising model uses a canonical ensemble (constant temperature) condition. In Q2R, all of the spins are flipped simultaneously, while in the probabilistic Ising model, only one randomly selected spin is flipped at a time. The spin-flipping decision is deterministic in Q2R while it is probabilistic in the Metropolis Ising model.

Both versions of the Ising model give many comparable results, although it should be pointed out that the Q2R program is not exactly equivalent to the Metropolis Ising program because the CA is not ergodic. We will not be concerned with those results here. Our focus will be on showing how the "checkerboard updating" scheme used in Q2R can be carried out.

10.1 THE Q2R MODEL

The Q2R model takes place on a two-dimensional $(n + 1) \times n$ lattice with *skewed* boundary conditions. The lattice sites have values of +1 or −1. The system evolves by simultaneously updating sites (known as *spin flipping*) based on their von Neumann neighborhoods, either until the system stops changing or for a specified number of times.

THE Q2R RULES

The constant-energy spin-flipping condition in the Q2R model is incorporated in the rule that states:

- A site is flipped only if it has an equal number of up spin and down spin neighbor sites.

THE Q2R ALGORITHM

1. The initial lattice configuration is given by

```
lat = Table[2 Random[Integer] - 1, {n + 1}, {n}]
```

2. The requirement that a spin flip not alter the energy of the system makes it necessary to simultaneously update only those spins that do not directly interact. To do this, the lattice is separated into two sublattices arranged like a checkerboard. Half of the sites are red and half are black such that no sites of the same color are nearest neighbors. On a lattice having an odd number of sites in each row, all of the odd-numbered sites form one sublattice, all of the even-numbered sites belong to the other sublattice, and the neighbors of sites in either sublattice are in the other sublattice.

The updating of the lattice is performed in two consecutive steps. First, the values of odd-numbered sites are updated using the values of their even-numbered nearest neighbor sites. Then, the values of even-numbered sites are updated using the updated values (determined in the previous step) of their odd-numbered nearest neighbor sites.

Each sublattice is updated using the following seven rules:

```
update[x_, -1, -1, 1, 1]  := -x
update[x_, -1, 1, -1, 1]  := -x
update[x_, -1, 1, 1, -1]  := -x
update[x_, 1, -1, -1, 1]  := -x
update[x_, 1, -1, 1, -1]  := -x
update[x_, 1, 1, -1, -1]  := -x
update[x_, _, _, _, _]  := x

Attributes[update] = Listable
```

The five arguments of the update function are the sites in the neighborhood of a site. They are (in order): the site (whose value is indicated by x), the nearest neighbor above the site, the nearest neighbor to the left of the site, the nearest neighbor to the right of the site, and the nearest neighbor below the site. The first six rules flip a site if there are equal numbers of up and down spin on neighboring sites. The last rule leaves the spin unchanged.

To use the update function, we create linear lists of the sites in each sublattice and of the various nearest neighbor sites for each sublattice. This can be demonstrated using a simple lattice.

```
In[1]:= (mat = Partition[Range[30], 5]) //MatrixForm
```

Out[1]//MatrixForm=

1	2	3	4	5
6	7	8	9	10
11	12	13	14	15
16	17	18	19	20
21	22	23	24	25
26	27	28	29	30

The following linear lists can be created from mat:

```
In[2]:= oddSites = Flatten[Partition[Flatten[mat], 1, 2]]
```
Out[2]= {1, 3, 5, 7, 9, 11, 13, 15, 17, 19, 21, 23, 25, 27, 29}

```
In[3]:= evenSites = Flatten[Partition[Rest[Flatten[mat]], 1, 2]]
```
Out[3]= {2, 4, 6, 8, 10, 12, 14, 16, 18, 20, 22, 24, 26, 28, 30}

```
In[4]:= ngbrsNorthOfOddSites = RotateLeft[evenSites, (-5 - 1)/2]
```
Out[4]= {26, 28, 30, 2, 4, 6, 8, 10, 12, 14, 16, 18, 20, 22, 24}

```
In[5]:= ngbrsWestOfOddSites = RotateLeft[evenSites, -1]
```
Out[5]= {30, 2, 4, 6, 8, 10, 12, 14, 16, 18, 20, 22, 24, 26, 28}

```
In[6]:= ngbrsEastOfOddSites= RotateLeft[evenSites, 0]
```
Out[6]= {2, 4, 6, 8, 10, 12, 14, 16, 18, 20, 22, 24, 26, 28, 30}

```
In[7]:= ngbrsSouthOfOddSites = RotateLeft[evenSites, (5 - 1)/2]
```
Out[7]= {6, 8, 10, 12, 14, 16, 18, 20, 22, 24, 26, 28, 30, 2, 4}

```
In[8]:= ngbrsNorthOfEvenSites = RotateLeft[oddSites, (-5 + 1)/2]
```
Out[8]= {27, 29, 1, 3, 5, 7, 9, 11, 13, 15, 17, 19, 21, 23, 25}

```
In[9]:= ngbrsWestOfEvenSites = RotateLeft[oddSites, 0]
```
Out[9]= {1, 3, 5, 7, 9, 11, 13, 15, 17, 19, 21, 23, 25, 27, 29}

```
In[10]:= ngbrsEastOfEvenSites = RotateLeft[oddSites, 1]
```
Out[10]= {3, 5, 7, 9, 11, 13, 15, 17, 19, 21, 23, 25, 27, 29, 1}

```
In[11]:= ngbrsSouthOfEvenSites = RotateLeft[oddSites, (5 + 1)/2]

Out[11]= {7, 9, 11, 13, 15, 17, 19, 21, 23, 25, 27, 29, 1, 3, 5}
```

These linear lists are used to carry out the update process. Here is how the odd-numbered sites are updated:

```
newOdd = update[oldOdd,
                        RotateLeft[oldEven, (-n - 1)/2],
                        RotateLeft[oldEven, -1],
                        RotateLeft[oldEven, 0],
                        RotateLeft[oldEven, (n - 1)/2]]
```

oldOdd and oldEven are the un-updated odd and even sublattices.
The updating of the even-numbered sites is carried out using,

```
newEven = update[oldEven,
                        RotateLeft[newOdd, (-n + 1)/2],
                        RotateLeft[newOdd, 0],
                        RotateLeft[newOdd, 1],
                        RotateLeft[newOdd, (n + 1)/2]]
```

where oldEven is the un-updated even sublattice and newOdd is the updated odd sublattice.

These two steps can be put together in a single function which returns a list of the new updated sublattices.

```
spinFlip[{oldOdd_, oldEven_}] :=
        Module[{newOdd, newEven},

            newOdd = update[oldOdd,
                    RotateLeft[oldEven, (-n - 1)/2],
                    RotateLeft[oldEven, -1],
                    RotateLeft[oldEven, 0],
                    RotateLeft[oldEven, (n - 1)/2]];

            newEven = update[oldEven,
                    RotateLeft[newOdd, (-n + 1)/2],
                    RotateLeft[newOdd, 0],
                    RotateLeft[newOdd, 1],
                    RotateLeft[newOdd, (n + 1)/2]];

            {newOdd, newEven}
        ]
```

3. The updating procedure is repeatedly performed using the NestList function with the ordered pair consisting of the sublattices.

```
sublatOdd = Flatten[Partition[Flatten[lat], 1, 2]];
sublatEven = Flatten[Partition[Rest[Flatten[lat]], 1, 2]];
NestList[spinFlip, {subLatOdd, subLatEven}, m]
```

4. In step **3**, we did not bother to reconstruct the lattice at each time step from its constituent sublattices. To do this (so that we can look at the graphics of the lattice), the sublattices are brought together.

```
Map[Partition[Flatten[Transpose[#]], n]&,
        NestList[spinFlip, {subLatOdd, subLatEven}, m]]
```

Now, we'll put all of this code together into a program.

THE Q2R PROGRAM

```
In[1]:= IsingCA[n_?OddQ, m_] :=
        Module[{lat, subLatOdd, subLatEven, update, spinFlip},
          lat = Table[2 Random[Integer] - 1, {n+1}, {n}];
          subLatOdd = Flatten[Partition[Flatten[lat], 1, 2]];
          subLatEven = Flatten[Partition[Rest[Flatten[lat]],1,2]];
          update[x_, -1, -1, 1, 1] := -x;
          update[x_, -1, 1, -1, 1] := -x;
          update[x_, -1, 1, 1, -1] := -x;
          update[x_, 1, -1, -1, 1] := -x;
          update[x_, 1, -1, 1, -1] := -x;
          update[x_, 1, 1, -1, -1] := -x;
          update[x_, _, _, _, _] := x;
          Attributes[update] = Listable;

          spinFlip[{oldOdd_, oldEven_}] :=
                Module[{newOdd, newEven},
                  newOdd = update[oldOdd,
                              RotateLeft[oldEven,(-n-1)/2],
                              RotateLeft[oldEven,-1],
                              RotateLeft[oldEven,0],
                              RotateLeft[oldEven,(n-1)/2]];
                  newEven = update[oldEven,
                              RotateLeft[newOdd,(-n+1)/2],
                              RotateLeft[newOdd,0],
                              RotateLeft[newOdd,1],
                              RotateLeft[newOdd,(n+1)/2]];
                  {newOdd, newEven}];
              Map[Partition[Flatten[Transpose[#]], n]&,
                NestList[spinFlip, {subLatOdd, subLatEven}, m]]
        ]
```

▇10.2▇ COMPUTER SIMULATION PROJECTS

1. It is rather intriguing that the Ising spin system can be implemented in both a probabilistic model and a deterministic model. Consider the scientific-philosophical implications if for any natural stochastic process, a deterministic cellular automaton can be written which can simulate it.

CHAPTER 11

Excitable Media

INTRODUCTION

Slime mold growth, star formation in spiral disk galaxies, cardiac tissue contraction, diffusion-reaction chemical systems and infectious disease epidemics would seem to be quite dissimilar systems. Yet, in each of these cases, various spatially distributed patterns, such as concentric and spiral wave patterns, are spontaneously formed. The underlying cause of the formation of these self-organized, self-propagating structures is that these are *excitable media*, consisting of spatially distributed elements which can become excited as a result of interacting with neighboring elements, subsequently returning incrementally to the quiescent state in which they are again receptive to being excited. The *excited-refractory-receptive* cycle that characterizes these systems can be modeled using multi-state cellular automata.

The three cellular automata models of excitable media described here (Greenberg-Hastings, cyclic space, and the hodgepodge CAs) differ in the way the recovery process depends on the interaction between neighboring sites. We will first develop all of the programs and then discuss their graphics.

11.1 NEURON EXCITATION

Electrical activity in biological systems can exhibit excitable media behavior. One example occurs in the contraction of cardiac tissue and is relevant to the phenomenon of sudden death, in which the heart muscle starts to fibrillate and then stops. This behavior, known as *malignant arrhythmia*, often occurs in the end stages of congestive heart failure, and sometimes in apparently healthy individuals during or following vigorous activity.

The two-dimensional CA, known as the Greenberg-Hastings CA (after its authors), has been used to model neuron excitation and recovery in a network of neurons. It takes place on a two-dimensional square lattice with periodic boundary conditions. Lattice site values range from 1 to r. Using neuro-physiological terminology, a neuron (site) with value 1 is said to be *excited*, a neuron with value r is said to be *rested*, and a neuron with an intermediate value is said to be in a *state of recovery.* The CA evolves by updating lattice sites a specified number of times, based on their von Neumann neighborhoods.

THE NEURON EXCITATION RULES

In the neuron excitation-recovery process, an excited neuron goes through a sequence of recovery states until it is rested. A rested neuron becomes excited if at least one nearest neighbor neuron is excited. The three update rules corresponding to this process are:

- A rested neuron (a neuron having value r) with at least one excited nearest neighbor neuron (a neuron having value 1) becomes excited (the neuron's value changes from r to 1).
- A rested neuron (a neuron having value r) with no excited nearest neighbor neurons remains rested (the neuron's value remains r).
- A neuron that is in a recovery state (a neuron having a value between 1 and $r-1$) goes to the next recovery state (the neuron's value increases by 1).

THE NEURON EXCITATION ALGORITHM

1. The initial configuration is an $s \times s$ lattice whose sites are randomly chosen integers between 1 and r.

```
neuronNet = Table[Random[Integer, {1, r}], {s}, {s}].
```

2. The update rules are implemented in a straightforward manner using pattern-matching:

```
neuron[a_, b_, r, d_, e_] := 1 /; MatchQ[1, a | b | d | e]
neuron[a_, b_, r, d_, e_] := r
neuron[_, _, c_, _, _] := c + 1

Attributes[neuron] = Listable
```

Using the above notation, the five arguments of neuron are described as follows: a is the value of the top (north) nearest neighbor cell; b is the value of the right (east) nearest neighbor cell; c is the value of the center cell; d is the value of the bottom (south) nearest neighbor cell; e is the value of the left (west) nearest neighbor cell.

The first rule says that a lattice site having a value r is updated to have a value of 1 if the value of any of its nearest neighbors is 1. The second rule says that a lattice site having a value r remains at r. The third rule says that a lattice site having a value c is incremented by 1.

The three neuron functions defined here are used as follows:

A. A cell in state *r* is updated with the first rewrite rule if its condition is met.

B. A cell in state *r* which does not meet the condition for using the first rewrite rule is updated with the second rewrite rule (the third rewrite rule is never used for a cell in state *r*).

C. A cell in state *c* is always updated with the third rewrite rule. Since the third rule applies only when the first two rules do not, it isn't necessary to specify the *c* < *r* condition in the third rule.

By making the update rule `Listable`, it can be applied directly to the five matrices whose corresponding elements are the values of a cell and its nearest neighbor cells.

3. The lattice is updated *t* times (*i.e.*, the system evolves over *t* time steps) using an anonymous function inside `NestList`.

```
NestList[
        (neuron[RotateRight[#], Map[RotateLeft, #], #,
              RotateLeft[#], Map[RotateRight, #]])&,
        neuronNet, t]
```

Here, # represents the lattice configuration and the five arguments to neuron are the matrices of sites and the nearest neighbors north, east, south, and west.

The program for neuron activation and relaxation is written from its constituents in the following section.

THE NEURON EXCITATION PROGRAM

```
In[1]:= NeuronExcitation[s_Integer, r_Integer, t_Integer] :=
        Module[{neuronNet, neuron},
            neuronNet = Table[Random[Integer, {1,r}], {s},{s}];

            neuron[a_, b_, r, d_, e_] := 1 /;
                                    MatchQ[1, a | b | d | e];
            neuron[a_, b_, r, d_, e_] := r;
            neuron[_, _, c_, _, _] := c + 1;
            Attributes[neuron] = Listable;

            NestList[
                (neuron[RotateRight[#], Map[RotateLeft, #], #,
                      RotateLeft[#], Map[RotateRight, #]])&,
                neuronNet, t]
        ]
```

11.2 CYCLIC SPACE

In the Greenberg-Hasting CA, the recovery of a site proceeds independently of the state of neighboring sites. The cyclic space CA is a modest variant of that model, allowing the recovery process to proceed for a site only when a neighboring site is slightly more relaxed. The behavior of this model has been related to the changes that occur in a phase transition.

The CA takes place on a square lattice with periodic boundary conditions. Lattice sites have values between 0 and $n-1$. The CA evolves by updating the lattice sites a specified number of times based on their von Neumann neighborhoods.

CYCLIC SPACE RULES

The sites are simultaneously updated according to the following three rules:

- A cell in state $n-1$ goes to state 0 if any of its nearest neighbor cells are in state 0 (*i.e.*, a cell in state $n-1$ is "eaten" by a nearest neighbor cell in state 0).
- A cell in state c, $0 \leq c < n-1$, goes to state $c+1$ if any of its nearest neighbor cells is in state $c+1$.
- A cell in state c, $0 \leq c < n-1$, remains in state c if it has no nearest neighbor cells in state $c+1$.

Comparing this rule set to the rule set in the neuron CA, we see that in both systems intermediate cell states are incremented by one, although in the neuron CA, incrementation is inescapable, while in the cyclic space CA it is conditional on the state of neighboring sites.

Note: This rule set has been (inappropriately) referred to as "dog-eat-dog" because each state (k) has one state ($k+1$) that can affect it and one state ($k-1$) that it can affect.

THE CYCLIC SPACE ALGORITHM

1. The starting lattice configuration is an $s \times s$ square lattice, known as *debris*, whose sites have randomly chosen integers between 0 and $n-1$.

```
debris = Table[Random[Integer, {0, n-1}], {s}, {s}]
```

2. The first and second CA rules in the cyclic space CA are both similar to the first rule in the neuron CA while the third rule is similar to the second rule in the neuron CA.

```
cyclicSpace[a_, b_, (n - 1), d_, e_] :=
        0 /; MatchQ[0, a | b | d | e]
cyclicSpace[a_, b_, c_, d_, e_] :=
        (c + 1) /; MatchQ[(c + 1), a | b | d | e]
cyclicSpace[_, _, c_, _, _] := c

Attributes[cyclicSpace] = Listable
```

The five arguments of the cyclicSpace rule are the same as for the neuron rules: a is the value of the top (north) nearest neighbor cell, b is the value of the right (east) nearest neighbor cell, c is the value of the center cell, d is the value of the bottom (south) nearest neighbor cell, and e is the value of the left (west) nearest neighbor cell.

3. The time evolution of the cyclic space CA is computed in the same way as is the evolution of the neuron CA.

```
NestList[
        (cyclicSpace[RotateRight[#],
            Map[RotateLeft, #], #, RotateLeft[#],
            Map[RotateRight, #]])&,
        debris, t]
```

THE CYCLIC SPACE PROGRAM

The program for the cyclic space CA looks exactly like the neuron CA program, except that it uses its own rule set.

```
In[1]:= Phases[s_Integer, n_Integer, t_Integer] :=
        Module[{debris,cyclicSpace},
            debris = Table[Random[Integer, {0,n-1}], {s},{s}];

            cyclicSpace[a_, b_, (n - 1), d_, e_] :=
                    0 /; MatchQ[0, a | b | d | e];
            cyclicSpace[a_, b_, c_, d_, e_] := (c + 1) /;
                        MatchQ[(c + 1), a | b | d | e];
            cyclicSpace[_, _, c_, _, _] := c;

            Attributes[cyclicSpace] = Listable;

            NestList[
                (cyclicSpace[RotateRight[#], Map[RotateLeft, #],
                 #, RotateLeft[#], Map[RotateRight, #]])&,
                debris, t]]
```

11.3 THE HODGEPODGE MACHINE

Pattern development in chemical systems has been of interest since Alan Turing's work relating it to biological morphogenesis. The hodgepodge[2] machine (a hodgepodge is a heterogeneous mixture) models autocatalytic chemical reactions in which two or more compounds combine, dissociate, and recombine in the presence of a catalyst. One example is the combination of carbon monoxide and oxygen to form carbon dioxide while they are absorbed on the surface of dispersed palladium crystallites. Another example is the Belousov-Zhabotinsky reaction in which malonic acid is oxidized by potassium bromate in the presence of iron or cerium.

The hodgepodge machine consists of an $s \times s$ square lattice with periodic boundary conditions. Lattice site values range from 0 to r. The lattice is called the *cell matrix* and the lattice sites are called *cells*. Cells having a value of 0 are said to be *healthy*; cells having a value r are said to be *ill*; all other cells (*i.e.*, cells having a non-zero value less than the maximum value) are said to be *infected*. (The higher the value, the more infected the cell is.) Sites in the hodgepodge CA are updated based on their von Neumann neighborhoods.

THE HODGEPODGE RULES

At each time step, all of the cells are simultaneously updated according to three rules. These rules are described below using a mixture of built-in functions and words to make them clearer than can be done using words alone.

- If a cell is ill, it becomes healthy.
- If a cell is healthy, it becomes infected and its value changes to an integer value which is determined by:

$$\texttt{Min} \left[\texttt{r}, \texttt{Floor} \left[\frac{\textit{number of infected nearest neighbor cells}}{\texttt{k1}} \right] \right. $$
$$\left. + \texttt{Floor} \left[\frac{\textit{number of ill nearest neighbor cells}}{\texttt{k2}} \right] \right]$$

where $\texttt{k1}$ and $\texttt{k2}$ are constants. Note: The use of the $\texttt{Min}$ function ensures that the value of a cell doesn't exceed r.

- If a cell is infected, it becomes more infected and its value changes to:

$$\texttt{Min} \left[\texttt{r}, \texttt{g} + \texttt{Floor} \left[\frac{\textit{sum of neighborhood cell values}}{\textit{number of infected neighborhood cells}} \right] \right]$$

where $\texttt{g}$ is is an integer known as the *speed of infection* constant.

---

[2]The word "hodgepodge" derives from the French word "hochepot" meaning a complex ingredient mix, esp. mutton stew.

THE HODGEPODGE ALGORITHM

1. The initial configuration consists of cells with randomly chosen values between 0 and r.

```
initconfig = Table[Random[Integer, {0, r}]
```

2. The three rules used to update the system are:

```
sick[r, _, _, _]  := 0
sick[0, x_, y_, _] := Min[r, Floor[y/k1] + Floor[x/k2]]
sick[c_, x_, y_, z_] :=
       Min[r, g + Floor[(c + (r x) + z)/(y + 1)]]
Attributes[sick] = Listable
```

The four arguments of the `sick` rewrite rules are the value of a cell, the number of ill nearest neighbors to the cell, the number of infected nearest neighbors to the cell, and the sum of the values of the infected nearest neighbors to the cell. The first rule *heals* a sick cell. The second rule *infects* a healthy cell. The third rule makes an infected cell *more infected*.

The three terms in the numerator on the right-hand side of the third rule are the value of the infected cell, `c`, the sum of the values of its ill nearest neighbors, `(x r)`, and the sum of the values of its infected nearest neighbors, `z`. Thus, the numerator, `(c + (x r) + z)`, is the sum of the values of all of the cells in the neighborhood while the denominator, `(1 + y)`, is 1 plus the number of infected nearest neighbors to the site (*i.e.*, the denominator is the total number of infected cells in the neighborhood of an infected cell).

3. The lattice is updated by applying an anonymous function to the initial lattice configuration a specified number of times. This is done with `FixedPointList`, with a third argument to stop the computation after a number of steps if it doesn't stop on its own.

```
FixedPointList[(sick[#,
              ngbrsVon[Floor[N[#/r]]],
              ngbrsVon[Sign[Mod[#, r]]],
              ngbrsVon[Mod[#, r]]])&,
           initconfig, t]
```

The computations of the quantities in the anonymous function are done using the `ngbrsVon` function. This function takes a matrix and returns a matrix whose elements are the sum of the values of the nearest neighbor sites of the corresponding sites in the matrix.

```
ngbrsVon[mat_] :=
    Apply[Plus, Map[RotateRight[mat, #]&,
                    {{1,0},{-1,0},{0,1},{0,-1}}]]
```

The arguments to the `ngbrsVon` function in the `sick` rules use the matrix of the system, #.

`Floor[N[#/r]]` is a Boolean matrix where 1 indicates an ill cell and 0 indicates an infected or healthy cell, so that `ngbrsVon[Floor[N[#/r]]]` is a matrix whose elements are the number of ill nearest neighbor cells for the corresponding cells in the # matrix.

`Sign[Mod[#, r]]` is a Boolean matrix where 1 indicates an infected cell and 0 indicates an ill or healthy cell, so that `ngbrsVon[Sign[Mod[#, r]]]` is a matrix whose elements are the number of infected nearest neighbor cells for the corresponding cells in the # matrix.

`Mod[#, r]` is a matrix of values ranging from 0 to $r - 1$, where 0 indicates an ill or healthy cell and other values indicate the degree of infection of an infected cell, so that `ngbrsVon[Mod[#, r]]` is a matrix whose elements are the sums of the values of the infected nearest neighbor cells for the corresponding cells in the # matrix.

Putting together all of the pieces described above, we can write the hodge-podge program.

THE HODGEPODGE PROGRAM

```
In[1]:= Hodgepodge[r_Integer, s_Integer, k1_, k2_,
                    g_Integer, t_Integer] :=
        Module[{initconfig, ngbrsVon, sick},
            initconfig = Table[Random[Integer,{0,r}], {s},{s}];
            ngbrsVon[mat_] :=
                Apply[Plus, Map[RotateRight[mat, #]&,
                                {{1,0},{-1,0},{0,1},{0,-1}}]];
            sick[r, _, _, _]  := 0;
            sick[0, x_, y_, _] := Min[r, Floor[y/k1] +
                                            Floor[x/k2]];
            sick[c_, x_, y_, z_] :=
                    Min[r, g + Floor[(c + (r x) + z)/(y + 1)]];
            Attributes[sick] = Listable;

            FixedPointList[(sick[#, ngbrsVon[Floor[N[#/r]]],
                            ngbrsVon[Sign[Mod[#, r]]],
                            ngbrsVon[Mod[#, r]]])&,
                        initconfig, t]]
```

11.4 GRAPHICAL OUTPUT OF EXCITABLE MEDIA MODELS

Graphical output of excitable media CAs can be created using a `RasterArray` with colors assigned according to the value of a site.

```
In[1]:= ShowExcitation[list_, opts___]:=
            Module[{coloring, r = Max[list]},
               coloring = Thread[Range[0, r] ->
                                   Map[Hue, Table[Random[], {r+1}]]];
               Show[Graphics[RasterArray[list /. coloring],
                  AspectRatio -> Automatic,
                  opts]]]
```

In the next few pages, we will display and describe some simulations for the models in this chapter. The CD-ROM contains additional functions for creating animations of these models in addition to the animations themselves and numerous single-frame images.

NEURON ACTION

A neuron activation process using a 100×100 square lattice having site values ranging from 1 to 16 is shown below. Each lattice site can assume one of 16 values, as represented by 16 different colors (or gray scales). The first frame shows the initial (random) configuration, with successive frames showing additional time steps.

```
In[2]:= SeedRandom[0]
        ShowExcitation[NeuronExcitation[100, 16, 0]]
```

Out[2]= -Graphics-

In[3]:= **SeedRandom[0]**
ShowExcitation[NeuronExcitation[100, 16, 10]]

Out[3]= -Graphics-

In[4]:= **SeedRandom[0]**
ShowExcitation[NeuronExcitation[100, 16, 25]]

Out[4]= -Graphics-

In[5]:= **SeedRandom[0]**
       ShowExcitation[NeuronExcitation[100, 16, 50]]

Out[5]= –Graphics–

In[6]:= **SeedRandom[0]**
       ShowExcitation[NeuronExcitation[100, 16, 100]]

Out[6]= –Graphics–

CYCLIC SPACE

The cyclic space model is run on a 256×256 grid where each site takes on one of 14 values. The cyclic space CA produces a succession of four patterns, or phases:

1. The initial *debris* of randomly distributed colors:

```
In[7]:= SeedRandom[4];
        ShowExcitation[Phases[256, 14, 1]];
```

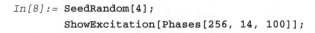

2. Compact regions of solid colors, known as *droplets*, spread out until they fill the space:

```
In[8]:= SeedRandom[4];
        ShowExcitation[Phases[256, 14, 100]];
```

3. Angular spirals, called crystalline *defects*:

```
In[9]:= SeedRandom[4];
        ShowExcitation[Phases[256, 14, 200]]
```

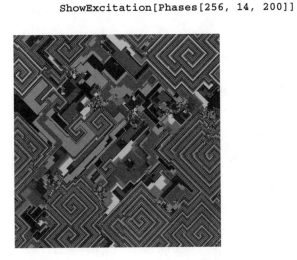

4. A few spirals, known as *demons*, that finally survive after the spirals grow and compete for space:

```
In[10]:= SeedRandom[4];
         ShowExcitation[Phases[256, 14, 500]];
```

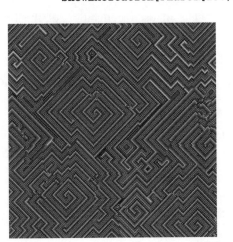

HODGEPODGE MACHINE

The hodgepodge machine produces four distinct regimes of wave patterns, appearing at increasingly higher values of the infection constant g, which resemble those shown above for the cyclic space CA (also see the spiral CA described in the fourth computer simulation project below).

1. Waves traveling (or meandering) a short distance and then dying out
2. Waves propagating in circular bands of differing widths
3. More-or-less uniform circular wave bands spreading out
4. Waves spreading in a spiral pattern

These four behavior regimes are also seen in plots of the fraction of infected cells over time for different values of g (see the first computer simulation project below).

The operation of the hodgepodge machine can be related to the microscopic details of the chemical reaction described on page 124 by giving a physical interpretation of the local interactions between the sites in the CA. The following correspondences between the CA and heterogeneous catalysis has been suggested: the cells to the catalyst particles, the degree of infection to the extent of coverage of the particle surfaces, and the local interactions to the transfer of heat and/or reactants between particles.

11.5 COMPUTER SIMULATION PROJECTS

1. A list of the fraction of infected cells in each generation of the hodgepodge machine can be created using the following function:

```
fractInfected =
  Map[(N[Apply[Plus,
        Flatten[Sign[#] - Floor[N[#/r]]]]]/s^2)&, reaction]
```

where `reaction` is the list of lattice configurations produced by running the Hodgepodge program. A plot of the number of infected cells versus the number of time steps can then be displayed.

```
ListPlot[fractInfected,
   PlotRange -> {Min[fractInfected], Max[fractInfected]},
   PlotJoined -> True,
   DefaultFont -> {"Times",12},
   PlotLabel -> FontForm["Fraction of Infected Cells",
                          {"Times",12}]
   ]
```

The fraction of infected cells over time can be examined for various values of the infection speed constant. For small values of g, all of the cells rapidly become healthy and then remain healthy. For increasingly higher values of g, four types of behavior, which are the counterparts to the four types of wave patterns, are seen:

A. Most of the cells are infected most of the time, while healthy cells occur irregularly and randomly.

B. There are plateaus in the number of infected cells separated by substantial spikes of healthy cells.

C. The incidence of infection regularly alternates between the infection of almost all and almost no cells.

D. The fraction of infected cells fluctuates regularly about a level of approximately 75 percent.

Create `ListPlots` of the fraction of infected cells in each generation of the hodgepodge machine that show these four regimes of behavior.

2. Dewdney studied the phases in the cyclic space CA and found that a loop structure apparently plays a role in droplet enlargement. A loop is defined as a closed chain consisting of cells having two nearest neighbor cells whose states differ from the cell's state by 1, 0, or −1 (when the sum of the state differences along the loop is non-zero, it is a defect).

   Modify the debris lattice in the cyclic space program to use a loop as a seed and observe what happens when the program is run.

3. One question about the behavior of the cyclic space CA concerns the relatively long persistence of each of the first three phases (or meta-states) before giving way to the next phase. Why does this happen?

   The answer is surprisingly simple. If there are n possible cell values, the probability that, for any cell, at least one of its nearest neighbors has a specific value (one more than the value of the cell) is given by one minus the probability that none of the nearest neighbor cells have the specific value. For a von Neumann neighborhood, this is given by:

```
incrementProbability[n_] := (1 - N[(n-1)/n]^4)
```

   So, for example, when there are 20 possible cell states, the increment probability is only 0.186. Hence, the long endurance of the intermediate states.

4. The spiral CA is a simplified variant of the Greenberg-Hastings CA, which rapidly produces the same spiral patterns.

   The spiral CA uses a two-dimensional square lattice with periodic boundary conditions. Lattice site values are −1, 0, or 1. Sites with value 1 are said to be excited while sites with value 0 are said to be in the resting state and sites with value −1 are said to be in the refractory state. The CA evolves by updating lattice sites a specified number of times, based on their von Neumann neighborhoods.

The evolution of the spiral CA proceeds according to the following update rules:

- A rested site (having value 0) with at least one excited nearest neighbor (having value 1) becomes excited (its value changes from 0 to 1).
- A rested site (having value 0) with no excited nearest neighbors remains rested (its value remains 0).
- A site that is in the refractory state (having value −1) goes to the rested state (its value changes from −1 to 0).
- An excited site (having value 1) goes to the refractory state (its value changes from 1 to −1).

Overall, these rules say that excited sites become refractory, rested sites with at least one excited nearest neighbors become excited, and all other sites become rested.

We can write the algorithm and its implementing *Mathematica* commands as follows:

The initial lattice configuration is a $2s \times 2s$ lattice. All the sites are rested, having value 0, except for a seed belt, consisting of:

A line of excited sites having value 1, running from the center to the left border of the lattice, and a line of refractory sites having value −1, running directly beneath the line of excited sites.

```
init = Fold[ReplacePart[#1, #2[[1]], #2[[2]]]&,
          Table[0, {2s}, {2s}],
          Join[Map[{1, {s, #}}&, Range[s]],
               Map[{-1, {s + 1, #}}&, Range[s]]]
       ]
```

The update rules are implemented in a straightforward manner:

```
spiral[a_, b_, 0, d_, e_] := Max[a, b, d, e, 0]
spiral[a_, b_, 1, d_, e_] := -1
spiral[_, _, _, _, _] := 0

Attributes[spiral] = Listable
```

The five arguments of `spiral` are, in order: the value of the top (north) nearest neighbor cell, the value of the right (east) nearest neighbor cell, the value of the center cell, the value of the bottom (south) nearest neighbor cell, and the value of the left (west) nearest neighbor cell.

The first rule says that a lattice site having value 0 is updated to value 1 if the value of any of its nearest neighbors is 1. The second rule says that a lattice site having value 1 is updated to value −1. The third rule says that a lattice site which does not satisfy either the first or second rule is updated to 0.

The program for the spiral CA is written as:

```
SpiralCA[s_Integer, t_Integer] :=
  Module[{init, spiral},
      init = Fold[ReplacePart[#1, #2[[1]], #2[[2]]]&,
                  Table[0, {2s}, {2s}],
                  Join[Map[{1, {s, #}}&, Range[s]],
                  Map[{-1, {s + 1, #}}&, Range[s]]]
              ] ;

      spiral[a_, b_, 0, d_, e_] := Max[a, b,  d, e, 0];
      spiral[a_, b_, 1, d_, e_] := -1;
      spiral[_, _, _, _, _] := 0;
      Attributes[spiral] = Listable;

      NestList[
          (spiral[RotateRight[#], Map[RotateLeft, #], #,
           RotateLeft[#], Map[RotateRight, #]])&,
          init, t]
  ]
```

Run the `SpiralCA` program to produce spiral patterns.

REFERENCES

* A. K. Dewdney. Computer Recreations: A cellular universe of debris, droplets, defects and demons. *Scientific American*, August (1989) 102–105.

* A. K. Dewdney. Computer Recreations: The hodgepodge machine makes waves. *Scientific American*, August (1988) 104–107.

R. Durret. Some new games for your computer. *Nonlinear Science Today*, 1 (1991) 1–7.

B. F. Madore and W. L. Freedman. Self-organizing structures. *American Scientist*, 75 (1987) 252–259.

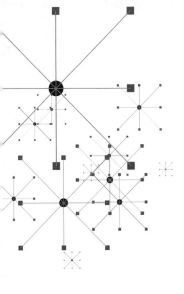

CHAPTER 12

Traffic

INTRODUCTION

Anyone who has been a passenger or driver in a car has an interest in traffic flow. When the traffic is moving smoothly and rapidly, it tends to be unnoticed, but as soon as the traffic slows down or becomes tied-up, the causes of the transition become a matter of speculation.

Traffic is actually a specific instance of a more general phenomenon known as the laminar-to-turbulent flow transition in a driven diffusive system with asymmetric exclusion. This system consists of particles, located on a lattice, having *hard core repulsion* so that no lattice site is occupied by more than one particle at a time. The particles have different probabilities of moving to the left and to the right, so that overall, there is a net current, or movement, of particles in a direction. This model has been used to study a wide variety of natural phenomena, ranging from vehicular traffic to surface roughening. The focus here will be modeling one-way traffic with one-dimensional cellular automata to examine the effect of stopping and passing on the flow of traffic as a function of the density of cars on the road.

12.1 THE ONE-LANE ROAD WITH CAR STOPPING

Traffic takes place on a one-dimensional lattice of length s, with periodic boundary conditions (so that the system has a ring geometry). The values of sites are either integers representing a car's speed ranging from 0 (a stopped car) to vmax (the speed limit), or the symbol e, where e represents an empty spot. The system evolves over a specified number of time steps by updating each lattice site, based on its value and the values of neighboring sites to its right.

THE ONE-LANE ALGORITHM

The sequence of steps **2** through **3** will be carried out a number of times. In implementing the step sequence, we'll use road to denote the configuration of the system in a given time step.

1. The initial `road` configuration consists of cars whose speeds and locations are both randomly distributed. This is created by first placing cars on the road and then determining their speeds.

```
road = Table[Floor[p + Random[]], {s}] *
           Table[Random[Integer, {1, vmax + 1}],{s}] /.
       {0 -> e, x_Integer -> (x - 1)}
```

Here `p` is the probability that a location on the road is occupied by a car, `e` represents an empty space on the road, and `s` represents the size of the lattice. The density of cars on the road is determined using

```
carDensity = N[Count[road, _Integer]/s]
```

The road has no exits or entrances and hence the car density is constant. In the limit of large `s`, the density will approach the value of `p`.

The movement of cars along the road is carried out by simultaneously updating all of the lattice sites according the following two-step updating procedure.

The order in which these two steps are carried out is all-important. Adjusting the car speed in step **2** before moving the car in step **3** ensures that no car is overtaken (smashed into) by the car behind it.

2. Car speed change: A car either slows down or speeds up, depending on its speed and the distance to the car ahead. If the distance, d, from a car to the car ahead is less than or equal to the velocity `vel` of the car, the car's speed slows down to `newVel = (d - 1)`; otherwise, the velocity of the car increases by 1 to `newVel = (vel + 1)`, up to the speed limit, `vmax`.

A car's velocity changes to the minimum of (a) one more than its present velocity, (b) one less than its distance from the car ahead, and (c) the speed limit. The list of updated car velocities, `newVels`, is calculated using

```
newVels = Map[(Min[#[[1]], #[[2]], vmax])&,
            Transpose[{vels + 1, distances - 1}]]
```

`vels` is the list of car velocities:

```
vels = DeleteCases[road, e]
```

The spacing between cars given by the function `distances`:

```
distances = Join[Rest[locs - RotateRight[locs]],
              {s - Last[locs] + First[locs]}]
```

Here, `locs` is the list of the current car locations:

```
locs = Flatten[Position[road, _Integer]]
```

The first argument to the `Join` function in `distances` is the list of the number of sites between the first car and the second car, the second car and the third car, and so on, up to the number of sites between the next-to-last car and the last car. The second argument to the `Join` function is a list containing a single element, which is the distance between the last car and the first car.

3. Car motion: A car moves to the right, a distance equal to the car's velocity, `newVel`.

   The new locations of cars is first determined, using:

```
newLocs = Map[(Mod[# - 1, s] + 1)&, (newVels + locs)]
```

Modular arithmetic is used to compute car movement, so that if a car "leaves" the system on the right, it reappears on the left in keeping with the periodic boundary conditions.

The cars are then placed in their new positions with their new velocities.

```
Fold[ReplacePart[#1, #2[[1]], #2[[2]]]&,
    Table[e, {s}], Transpose[{newVels, newLocs}]]
```

We can combine the code for performing the sequence of steps **2** through **3** into an anonymous function, using the symbol `y` to represent the `road` configuration.

```
followTheLeader =
  Function[y,
    vels = DeleteCases[y, e];
    locs =  Flatten[Position[y, _Integer]];
    distances =
      Join[Rest[locs - RotateRight[locs]],
        {s - Last[locs]] + First[locs]}];
    newVels = Map[(Min[#[[1]], #[[2]], vmax])&,
            Transpose[{vels + 1, distances - 1}]];
    newLocs = Map[(Mod[# - 1, s] + 1)&,
            (newVels + locs)];
    Fold[ReplacePart[#1, #2[[1]], #2[[2]]]&,
      Table[e, {s}],
      Transpose[{newVels, newLocs}]]
    ]
```

4. The evolution of the traffic proceeds by performing *t* updates to the system.

```
NestList[followTheLeader, road, t]
```

We're now ready to construct the entire program.

THE ONE-LANE PROGRAM

```
In[1]:= OneLane[s_, p_, vmax_, t_] :=
         Module[{emptyRoad, road, carDensity, followTheLeader},
           emptyRoad = Table[e, {s}];
           road = Table[Floor[p + Random[]],{s}] *
                  Table[Random[Integer,{1, vmax + 1}], {s}] /.
             {0 -> e, x_Integer -> (x - 1)};
           carDensity = N[Count[road, _Integer]/s];
           followTheLeader =
             Function[y,
               vels = DeleteCases[y,e];
               locs = Complement[Range[s], Flatten[Position[y,e]]];
               distances = Join[Rest[locs - RotateRight[locs]],
                               {s - Last[locs] + First[locs]}];
               newVels = Map[(Min[#[[1]], #[[2]], vmax])&,
                             Transpose[{vels + 1, distances - 1}]];
               newLocs = Map[(Mod[# - 1, s] + 1)&,
                             (newVels + locs)];
               Fold[ReplacePart[#1, #2[[1]], #2[[2]]]&,
                    emptyRoad, Transpose[{newVels, newLocs}]]];
           NestList[followTheLeader, road, t]
         ]
```

RUNNING THE ONE-LANE PROGRAM

A graphic of the evolution of the traffic model with the road shown in black
(Hue[0,0,0]) and the cars represented as hues, with different colors representing different velocities, can be created with a RasterArray.

```
In[2]:= ShowTraffic[list_List, opts___] :=
         Module[{vmax = Max[list/. e -> -1]},
           Show[Graphics[RasterArray[Reverse[list] /.
             Join[{e -> Hue[0, 0, 0]},
               Thread[Range[0, vmax] ->
                   (Map[Hue, Table[Random[],{vmax + 1}]])])]
           ]]],
           opts, AspectRatio -> Automatic]]
```

Using the ShowTraffic function with the OneLane program, we observe
two different regimes (phases) of behavior as a function of car density. This is shown
in the two figures below for a road of length 1000 and a maximum velocity of 10,
over 500 time steps, with car densities of 0.15 and 0.25, respectively.

In[3]:= **ShowTraffic[OneLane[1000, 0.15, 10, 500]]**

Out[3]= -Graphics-

In[4]:= **ShowTraffic[OneLane[1000, 0.25, 10, 500]]**

Out[4]= -Graphics-

The figures are rather abstract looking but can be interpreted in terms of *kinematic waves* which move at different speeds and meet to produce *shock waves*. The kinematic waves are created as cars adjust their speed to the cars ahead, creating waves of constant flow (packs of fast-moving, widely spaced cars and slow-moving, closely spaced car caravans) which travel either in the same or opposite direction as the traffic is moving. A shock wave is created when a fast-moving car must brake suddenly to avoid running into a slow-moving car. The visual identification of traffic waves can be supplemented by plots of the car flow and mean speed as a function of car concentration. We will look at this in the next traffic CA.

12.2 THE TWO-LANE, ONE-WAY ROAD WITH CAR PASSING

The TwoLane model is a two-state CA consisting of two interacting car lanes going in the same direction. The two states correspond to empty spaces and to cars. The interaction occurs when a car that is blocked by a car immediately ahead shifts to the adjacent position in the next lane (*i.e.*, switches lanes) if possible.

The model takes place on two one-dimensional lattices of length *s*, each with periodic boundary conditions. The sites have values 0 (an empty space) or 1 (a car). The system evolves over a specified number of time steps by updating each lattice site, based on its value and the values of neighboring sites to the right, above, and below.

In contrast to the previous CA, there is no distribution of car speeds. In a given time step, a car will either stop (in which case it remains in its current location), or it will move one space forward or sideways.

THE TWO-LANE ALGORITHM

1. The initial highway configuration consists of two one-dimensional lattices whose sites have probabilities p of being occupied by a car (a site having a value 1) and (1 − p) of being empty (a site having a value 0).

   ```
   highway = Table[Floor[p + Random[]],{2}, {s}]
   ```

   The density of cars on the highway is given by

   ```
   totalcars = Apply[Plus, Flatten[highway]]
   ```

   and

   ```
   carDensity = N[totalcars/(2s)]
   ```

2. The movement of cars along the highway is carried out by simultaneously updating all of the lattice sites in the following two step sequence.

   A. At *t* + 0.5, a car in the right lane moves ahead, shifts to the left lane, or stops. The car moves ahead if there is no car in front. The car shifts to the left lane if there is a car ahead of it but no car next to it in the left lane. The car stops if there are cars ahead and next to it.

   B. At *t* + 1, a car in the left lane moves ahead, shifts to the right lane, or stops. The car moves ahead if there is no car in front. The car shifts to the right lane if there is a car ahead of it but no car next to it in the right lane. The car stops if there are cars ahead and next to it.

When the system is updated during each half-step, each site in the lane and the site in the adjacent lane are simultaneously updated. The rules sets for updating the right lane and left lane are identical except that the order of elements in ordered pairs on the right-hand side of the rules is reversed.

```
ruleR[1, 0, _, x_] := {x, 1}
ruleR[_, 1, 1, 0] := {1, 0}
ruleR[_, 1, 0, x_] := {x, 0}
ruleR[_, y_, _, x_] := {x, y}
Attributes[ruleL] = Listable

ruleL[1, 0, _, x_] := {1, x}
ruleL[_, 1, 1, 0] := {0, 1}
ruleL[_, 1, 0, x_] := {0, x}
ruleL[_, y_, _, x_] := {y, x}
Attributes[ruleL] = Listable
```

The arguments on the left-hand side of the rules are: the nearest neighbor behind the site, the site, the nearest neighbor ahead of the site, and the nearest neighbor next to the site. The ordered pair on the right-hand side of the rule consists of the value of the nearest neighbor adjacent to the site and the value of the site itself.

In the first rule of the set `ruleR` (`ruleL`), the empty site in the right (left) lane becomes occupied and the adjacent site in the left (right) lane remains unchanged. This corresponds to a car behind moving into an empty site.

In the second rule, the occupied site in the right (left) lane becomes empty and the empty adjacent site in the left (right) lane becomes occupied. This corresponds to lane switching.

In the third rule, the occupied site in the right (left) lane becomes empty and the adjacent site in the left (right) lane remains unchanged. This corresponds to a car moving into an empty site ahead.

The fourth rule corresponds to no change. This corresponds to all other possible situations; *e.g.*, a car being blocked both ahead and on the side (and therefore stopping) or an empty site with neither a car behind nor a blocked car next to it (and therefore remaining empty).

In step **2.A**, the update of the right lane is performed using the `ruleR` rules with the following function:

```
laneRupdate[lis_] :=
    Transpose[ ruleR[RotateLeft[lis[[2]], -1], lis[[2]],
                RotateLeft[lis[[2]], 1], lis[[1]]] ]
```

The application of `ruleR` produces s ordered pairs (representing the updating of adjacent highway sites) and `Transpose` is then used to reconstruct the road.

In step **2.B**, the update of the left lane is performed using the `ruleL` rules with the function:

```
laneLupdate[lis_] :=
    Transpose[ ruleL[RotateLeft[lis[[1]], -1], lis[[1]],
                      RotateLeft[lis[[1]], 1], lis[[2]]] ]
```

The overall time step update consisting of the consecutive application of steps **2.A–B**, is performed with a nested function call:

```
stopAndGo[lane_] := laneLupdate[laneRupdate[lane]]
```

In both parts of the update step, the number of cars is conserved.

3. The highway system evolves through *t* time steps using `stopAndGo` with the initial configuration given by `hghwy`.

```
NestList[stopAndGo, hghwy, t]
```

We can now construct the program from these parts.

The Two-Lane Program

```
In[1]:= TwoLane[s_, t_, p_] :=
            Module[{highway, totalcars, carDensity, ruleR,
                    count = 0, velLis = {}, ruleL, laneRupdate,
                    laneLupdate, stopAndGo},

                highway = Table[Floor[p + Random[]],{2}, {s}];
                totalcars = Apply[Plus, Flatten[highway]];
                carDensity = N[totalcars/(2s)];
                ruleR[1, 0, _, x_] := (count++; {x, 1});
                ruleR[_, 1, 1, 0] := {1, 0};
                ruleR[_, 1, 0, x_] := {x, 0};
                ruleR[_, y_, _, x_] := {x, y};
                Attributes[ruleR] = Listable;
```

```
ruleL[1, 0, _, x_] := (count++; {1, x});
ruleL[_, 1, 1, 0] := {0, 1};
ruleL[_, 1, 0, x_] := {0, x};
ruleL[_, y_, _, x_] := {y, x};
Attributes[ruleL] = Listable;

laneRupdate[lis_] :=
    Transpose[ ruleR[RotateLeft[lis[[2]],-1], lis[[2]],
            RotateLeft[lis[[2]], 1], lis[[1]]] ];

laneLupdate[lis_] :=
    Transpose[ ruleL[RotateLeft[lis[[1]],-1], lis[[1]],
            RotateLeft[lis[[1]], 1], lis[[2]]] ];

stopAndGo[lane_] := Module[{},
            AppendTo[velLis, count];
            count = 0;
            laneLupdate[laneRupdate[lane]]];

NestList[stopAndGo, hghwy, t];
velLis = Join[velLis, {count}];
MeanVelocityLis = N[velLis/totalcars]
]
```

In constructing the program from its constituent parts, we have introduced a counter, cleverly called count, so that we can create the so-called "fundamental diagram," the flow of cars per time step versus car density. The *mean velocity* is defined as the number of cars moving ahead in a time step divided by the total number of cars. In order to determine this quantity, we attach count to the first rule. At the start of each time step, the value of count from the previous time step is placed in a list of counts for each time step, count is reset to zero, and count is then incremented each time the first rule is applied during the time step.

RUNNING THE TWO-LANE PROGRAM

We can create a plot of the mean velocity versus the car density using values of velLis obtained from running the TwoLane program for specified values of s, p and t. The curve in the graph on the next page was calculated for a road of length 100 on the 200th time step.

The figure shows that the mean velocity undergoes a phase transition between low-density and high-density behaviors. The mean velocity is 1 from zero density until a critical density is reached and a "traffic jam" transition occurs, after which it then monotonically decreases, reaching 0 at a density of one. The figure also shows that the mean lane switching (see the first computer simulation project below) is zero until just above the phase transition point, after which it rises to a maximum and then falls back to zero.

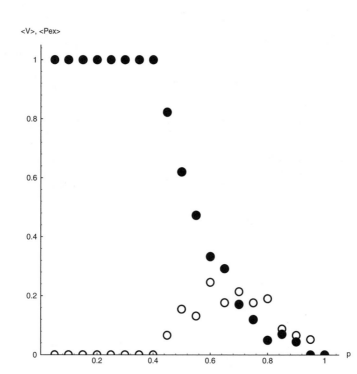

<V>, <Pex>

12.3 ■ COMPUTER SIMULATION PROJECTS

1. Modify the TwoLane program to calculate the mean number of lane changes per time step and reproduce the behavior of the open circles in the last figure.
2. Modify the TwoLane program for the limiting case when lane switching is not allowed. Draw the fundamental curve for this case and compare it with the corresponding curve in the last figure.
3. The presence of an obstacle (a traffic accident or road work) can be easily incorporated into the TwoLane program. This requires only the following code changes in the program (use k = 1 for an accident and k = 0 for road work).

```
hghwy =
ReplacePart[Table[Floor[p + Random[]],{2}, {s}],
            c,
            {Random[Integer,{1, 2}], Random[Integer,{1, s}]}]


ruleR[_, 1, (1 | c), 0] := {1, 0}
ruleL[_, 1, (1 | c), 0] := {0, 1}


totalcars = Apply[Plus, Flatten[hghwy /. c -> k]];
```

Explain how these changes work, then write the program for one-way, two-lane traffic in the presence of obstacles and determine the effect of obstacles on the fundamental diagram.

REFERENCES

* J. Walker. The Amateur Scientist: How to analyze the shock waves that sweep through expressway traffic. *Scientific American,* (August 1989) 98–101.

* B. Holmes. When shock waves hit traffic. *New Scientist,* 25 (June 1994) 36–40.

T. Nagatani. Self-organization and phase transition in a traffic flow model of two-lane roadway. *J. Phys. A* 26 (1993) L781–787.

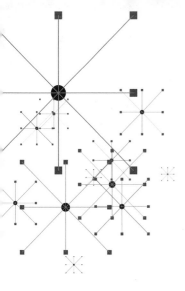

CHAPTER 13

Forest Fires

INTRODUCTION

It has been suggested that turbulent cascading, which is a manifestation of the SOC state, is the underlying cause of a wide variety of phenomena, including geological upheavals (*e.g.*, volcanic eruptions and earthquakes), species extinction during evolution, and fluid turbulence. Another system that displays this behavior is the forest fire cellular automaton of a forest preserve with clusters of trees (forests) separated by open spaces. In this chapter, we study this cellular automaton and look into such topics as forest distribution and size in forest preserves.

13.1 THE FOREST FIRE CA

The forest fire CA employs a one-dimensional lattice of length n, with periodic boundary conditions. Lattice sites may have a value of 0, 1, or 2, where 0 represents an empty site (or hole), 1 represents a healthy tree (or tree), and 2 represents a burning tree. A forest is a set of contiguous sites (*i.e.*, a connected segment) that have values 1 or 2. A forest preserve consists of forests separated by gaps (or deserts) which are clusters of connected holes. The system evolves for a specified number of time steps, in each of which entire forests of trees can catch fire and burn down and trees can sprout on individual, empty sites. Because trees sprout individually but burn down collectively, there is a separation in the time scales for the processes of deforestation and reforestation.

THE FOREST FIRE ALGORITHM

1. A forest preserve of length n, consisting of empty sites and tree sites can be created as follows:

```
forestPreserve = Table[Random[Integer], {n}]
```

2. All of the sites in the forest preserve are updated in each time step, according to the following sequence of steps:

A. Trees catch fire with probability f and empty sites sprout trees with probability p, according to a set of transformation rules that turn 0s into 1s with probability p, and 1s into 2s with probability f.

```
treeGrowIgnite = forestPreserve /.
                    {0 :> Floor[1 + p - Random[]],
                     1 :> Floor[2 + f - Random[]]}
```

B. All tree sites adjacent to an ignited tree site (*i.e.*, trees in the same forest) are ignited through repeated application of a set of transformation rules to the system.

```
forestIgnite = treeGrowIgnite //.
                    {{a___,  2, 1, b___} -> {a, 2, 2, b},
                     {a___,  1, 2, b___} -> {a, 2, 2, b},
                     {2, c___, 1} -> {2, c, 2},
                     {1, c___, 2} -> {2, c, 2}}
```

The first two transformation rules change sequences in the list having the forms ..., 1, 2, ..., or ..., 2, 1, ... to ..., 2, 2, ...; the last two transformation rules implement periodic boundary conditions thus treating trees at each end of the system as belonging to the same forest.

C. All ignited trees are burnt down using the transformation rule

```
forestNew = forestIgnite /. 2 -> 0
```

The sequence of steps **2.A–C** can be combined in an anonymous function that can be applied to any forest preserve configuration, where # represents the forest preserve configuration.

```
pyro =
(treeGrowIgnite = # /. {0 :> Floor[1 + p - Random[]],
                        1 :> Floor[2 + f - Random[]]};
 forestIgnite = treeGrowIgnite //.
                    {{2, c___, 1} -> {2, c, 2},
                     {1, c___, 2} -> {2, c, 2},
                     {a___, 2, 1, b___} -> {a, 2, 2, b},
                     {a___, 1, 2, b___} -> {a, 2, 2, b}};
 forestNew = forestIgnite /. 2 -> 0)&
```

3. Step **2** is repeated *m* times, using

```
NestList[pyro, forestPreserve, m]
```

These pieces of code can be combined into a program.

The Forest Fire Program

```
In[1]:= SmokeyTheBear[n_, p_, f_, m_] :=
            Module[{forestPreserve, pyro, treeGrowIgnite,
                forestIgnite, forestNew},
              forestPreserve = Table[Random[Integer], {n}];
              pyro = (treeGrowIgnite = #/.
                        {0 :> Floor[1 + p - Random[]],
                         1 :> Floor[2 + f - Random[]]};
                      forestIgnite = TreeGrowIgnite //.
                        {{2, c___, 1} -> {2, c, 2},
                         {1, c___, 2} -> {2, c, 2},
                         {a___, 2, 1, b___} -> {a, 2, 2, b},
                         {a___, 1, 2, b___} -> {a, 2, 2, b}};
                      forestNew = forestIgnite /. 2 -> 0)&;
              NestList[pyro, forestPreserve, m]
            ]
```

13.2 ■ Forest Size Distribution

It is useful to see the "forest through the trees." To identify the various forests in the forest preserve, a program that labels clusters on a one-dimensional lattice with periodic boundary conditions can be written, consisting of the following steps:

1. The sites in the lattice `lat` are updated using

```
clusterID[#, RotateLeft[#]]&[lat]
```

where

```
clusterID[1, 0] := i++
clusterID[1, 1] := i
clusterID[0, b_] := 0
Attributes[clusterID] = Listable
```

The two arguments of `clusterID` are the value of a site and the value of its right nearest neighbor. `clusterID` places an identifying label, i, on the sites in each cluster of occupied sites. The initial value of i is 1 and increases by 1 for each successive cluster.

2. After `clusterID` has been applied to the lattice, all of the clusters have been correctly labeled with one exception: Clusters occurring at the far ends of the lattice have been misidentified as different clusters when they should be treated as a single cluster. This can be corrected using an anonymous function:

```
        If[MatchQ[0, First[#] | Last[#]], #, # /.
             Last[#] -> 1]&
```

These code fragments can be combined into a program.

```
In[1]:= clusterLabel1D[lis_List] :=
            Module[{i = 1, clusterID, result},
                clusterID[1, 0] := i++;
                clusterID[1, 1] := i;
                clusterID[0, b_] := 0;
                Attributes[clusterID] = Listable;
                result = clusterID[#, RotateLeft[#]]&[lis];
                If[MatchQ[0, First[#] | Last[#]],
                   #,
                   # /. Last[#] -> 1]&[result]]
```

The use of the `clusterLabel1D` function can be illustrated with a simple forest preserve.

```
In[2]:= config = {1, 1, 0, 0, 1, 0, 1, 1, 1, 0, 1};
        clusterLabel1D[config]

Out[2]= {1, 1, 0, 0, 2, 0, 3, 3, 3, 0, 1}
```

The sizes of the various forests in the forest preserve can be determined using a modified version of the `clusterLabel1D` program:

```
In[3]:= TreeCount[lis_]:=
            Module[{counter, i = 1, forestSizes},
                counter[1, 0] := i;
                counter[1, 1] := (i++; 0);
                counter[0, 0] := 0;
                counter[0, 1] := (i = 1; 0);
                Attributes[counter] = Listable;
                forestSizes =
                   DeleteCases[counter[#,
                                    RotateLeft[#]]&[Join[lis,{0}]],0];
                If[MatchQ[lis, {1, ___, 1}],
                   Join[{First[#] + Last[#]}, Take[#, {2, -2}]],
                   #]&[forestSizes]]
```

The `counter` rule replaces the value of the right most tree in each forest with the number of trees in the forest, and the values of all the other trees in the forest with 0.

Here, `TreeCount` is used with a simple forest preserve `config`:

```
In[4]:= config = {1, 1, 0, 0, 1, 0, 1, 1, 1, 0, 1};
        forestSizes = TreeCount[config]

Out[4]= {3, 1, 3}
```

This last result indicates that the forests in `config` have 3, 1, and 3 trees, respectively.

While `forestSizes` gives the sizes of forests in the order in which they occur in the forest preserve, it does not give the sizes of the deserts lying between the forests. This can be determined using the following calculation, which indicates that `config` consists of a sequence of a three-tree forest, a two-site desert, a lone tree, a single empty site, a three-tree forest, and an empty site.

```
In[5]:= Map[({#, 1})&, config] //.
        {u___, {v_, r_}, {v_, s_}, w___} -> {u, {v, r + s}, w} /.
        {{y_, t_}, z___, {y_, u_}} -> {{y, t + u}, z}

Out[5]= {{1, 3}, {0, 2}, {1, 1}, {0, 1}, {1, 3}, {0, 1}}
```

Finally, the distribution of forest sizes can be computed using `forestSizes` with the `Frequency` function.

```
In[6]:= Frequency[x_List] := Map[{#, Count[x, #]}&, Union[x]]
```

```
In[7]:= Frequency[forestSizes]

Out[7]= {{1, 1}, {3, 2}}
```

This indicates that `config` contains one lone tree and two three-tree forests.

A graph of the forest size distribution can be made using `ListPlot`.

13.3 **COMPUTER SIMULATION PROJECTS**

1. Determine whether the distribution of forest sizes in the forest fire model follows exponential or power law behavior. Note: Power law correlations in space and time are indicative of SOC (see Chapters 7 and 9).

2. A two-dimensional version of the forest fire algorithm has recently been presented that appears to show a variety of interesting behaviors (see Drossal and Schwabl 1994). In this model, the sites of a square lattice with periodic boundary conditions are either occupied by trees (*i.e.*, have a value 1), or by burning trees (*i.e.*, have a value 2), or are empty (*i.e.*, have a value 0). During a time step, all of the sites are updated according to the following rules:

- A tree becomes a burning tree with probability $(1 - g)$ if at least one nearest neighbor is burning.
- A tree becomes a burning tree with probability $f(1 - g)$ if no nearest neighbor is burning.
- A burning tree becomes an empty site.
- An empty site becomes a tree with probability p.

There are three probability parameters: f is the lightning probability, g is the immunity, and p is the tree-growth probability. The following has been found for this model:

i. When $f = g = 0$, spiral-shaped fire fronts, reminiscent of excitable media (see Chapter 11), form for small values of p. (If you think of a tree as a *quiescent* state, a burning tree as an *excited* state, and the empty site as a *refractory* state, this behavior is not unexpected.)

ii. When $g = 0$, a self-organized critical state (see Chapter 9) occurs.

iii. When $f = 0$, a percolation-like phase transition (see Chapter 5) takes place at a critical value of g, which depends on p.

The spirals form when empty areas grow at the expense of forests as fire advances while forests grow into empty areas in the absence of fire, so that the fire fronts and tree fronts wind around one another. The self-organized critical state results when tree growth occurs more often than lightning and forests burn down faster than trees grow, so that fires of all sizes occur. The percolation-like phase transition happens when there is a zero fire density.

Implement this algorithm in a program and use it to create snapshots of the forest preserve showing these behaviors.

3. The average size of forests in a two-dimensional forest preserve is of interest. To determine this quantity, we first identify the forests in the preserve and then measure their sizes.

The labeling procedure for the two-dimensional lattice having periodic boundaries can be explained using a simple lattice as an example.

```
In[1]:= (lat = Table[Random[Integer],{4},{5}]) //MatrixForm

Out[1]//MatrixForm= 1   0   1   0   0
                    0   1   0   1   0
                    1   1   0   1   0
                    1   0   0   0   1
```

A. The lattice sites at the northwest (top left) corners of the clusters are first labeled.

```
clusterCornerID[RotateRight[lat, {1, 0}],
        RotateRight[lat, {0, 1}], lat]
```

where the transformations rules are give by

```
ClusterCornerID[1, 0, 0] := i++
ClusterCornerID[a_, __] := a
Attributes[ClusterCornerID] = Listable
```

The three arguments of `ClusterCornerID` are the value of a site, the value of the nearest neighbor above the site, and the value of the nearest neighbor to the left of the site.

Using `ClusterCornerID` with `lat`, we get

```
In[2]:= i = 2;
        ClusterCornerID[1, 0, 0]:= i++;
        ClusterCornerID[a_, __]:= a;
        Attributes[ClusterCornerID] = Listable;
        lat = {{1,0,1,0,0}, {0,1,0,1,0},
               {1,1,0,1,0}, {1,0,0,0,1}};
        (cornerLabels =
           ClusterCornerID[#,
                       RotateRight[#,{1,0}],
                       RotateRight[#,{0,1}]
                       ]&[lat]) //MatrixForm
```

```
Out[2]//MatrixForm= 1   0   2   0   0
                    0   3   0   4   0
                    5   1   0   1   0
                    1   0   0   0   6
```

In performing the above labeling, the initial value of i is taken to be 2 (and as a result, cluster corner sites were numbered 2, 3, ...) because the value 1 is used to identify cluster sites that are not corners.

B. Having labeled the sites at the corner of clusters, we next label the sites (those with value 1) that lie within clusters and we merge contiguous clusters. Both of these are accomplished with the following rules.

```
reLabel[0, ___] := 0
reLabel[a_, b_, c_, d_, e_] := Max[a, b, c, d, e]
Attributes[reLabel] = Listable
```

The five arguments of reLabel are the value of a site, the value of the nearest neighbor above the site, the value of the nearest neighbor to the left of the site, the value of the nearest neighbor beneath the site, and the value of the nearest neighbor to the right of the site. We use FixedPoint to apply reLabel repeatedly to cornerLabels (which is the result of applying ClusterCornerID to lat), until the cluster labels no longer change.

```
In[3]:= reLabel[0, ___] := 0;
        reLabel[a_, b_, c_, d_, e_] := Max[a, b, c, d, e];
        Attributes[reLabel] = Listable;
        FixedPoint[reLabel[#,
                           RotateRight[#,{1, 0}],
                           RotateRight[#,{0, 1}],
                           RotateRight[#,{-1,0}],
                           RotateRight[#,{0,-1}]]&,
                   cornerLabels] //MatrixForm
```

```
Out[3]//MatrixForm= 6   0   2   0   0
                    0   6   0   4   0
                    6   6   0   4   0
                    6   0   0   0   6
```

We can now combine these code fragments into a program.

```
ClusterLabel2D[lat_List]:=
  Module[{i=2, ClusterCornerID, cornerLabels, reLabel},
        ClusterCornerID[1,0, 0]:= i++;
        ClusterCornerID[a_, __]:= a;
        Attributes[ClusterCornerID] = Listable;

        cornerLabels =
           ClusterCornerID[#, RotateRight[#,{1,0}],
                           RotateRight[#,{0,1}]]&[lat];

        reLabel[0, ___] := 0;
        reLabel[a_, b_, c_, d_, e_] := Max[a, b, c, d, e];
        Attributes[reLabel] = Listable;

        FixedPoint[reLabel[#,
                           RotateRight[#,{1, 0}],
                           RotateRight[#,{0, 1}],
                           RotateRight[#,{-1,0}],
                           RotateRight[#,{0,-1}]]&,
                   cornerLabels]]
```

If desired, it is straightforward to relabel the identified clusters so that their numbering is sequential with no gaps (see the relabelrules2 function in the ClusterLabel program in Chapter 5).

While the `ClusterLabel2D` program is specifically designed for use with a lattice system having periodic boundary conditions, it can be used with a lattice having absorbing boundary conditions by "decorating" the lattice with rows and columns of 0s (the `Sandpile` model program in Chapter 9 shows how to do this using the `absorbBC` function).

Compare the relative speeds of the `ClusterLabel2D` program and the `Cluster-Label` program in Chapter 5 as functions of the lattice size.

REFERENCES

* Per Bak and Maya Paczuski. Why nature is complex. *Physics World*, (December 1993) 396–403.

B. Drossal and F. Schwabl. Self-organized critical forest fire model. *Phys. Rev. Lett.* 69 (1992) 1629–1632.

B. Drossal and F. Schwabl, *Physica A* 204 (1994) 212–229.

Maya Paczuski and Per Bak. Theory of the one-dimensional forest fire model. *Phys. Rev. E* 48 (1993) R3214–3216.

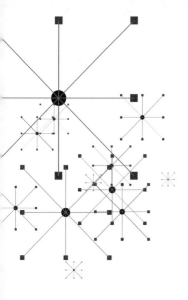

CHAPTER 14

Complexity

INTRODUCTION

Stephen Wolfram has made an extensive study of a type of one-dimensional cellular automata consisting of a row of lattice sites with periodic boundary conditions. The lattice sites in a Wolfram CA evolve according to deterministic, homogeneous (uniform), and local rules. These rules depend on the number of interacting lattice sites (*i.e.*, for a neighborhood containing nearest neighbor sites on each side of the site) and the number of values that a lattice site can have.

Wolfram found that these CAs evolve into patterns falling into one of four general classes:

- homogeneous state (analogous to limit points)
- periodic structures (analogous to limit cycles)
- chaotic patterns (analogous to strange attractors)
- complex, localized structures (possibly analogous to universal computers)

The analogies of the CA behaviors in the first three classes to the behaviors of physical systems (which are noted in parentheses above) is quite interesting. Even more intriguing is the behavior in the fourth class, where there is no simple algorithm for predicting behavior and it is necessary to actually run the CA simulation to determine the outcome.

We'll develop a program for generating Wolfram cellular automata. We'll first describe how to create the rule sets used to update lattice sites and the numbering scheme developed by Wolfram for these rule sets. After that, we'll present the algorithm for creating the CAs and then finally show some examples from these general classes.

14.1 THE WOLFRAM CA MODEL

Lattice sites in a Wolfram CA are updated using a rule set. A rule set consists of a list of update rules based on specific values (ranging from 0 to $(k - 1)$) of each of the $(2r + 1)$ sites in a neighborhood.

For example, the simplest nontrivial Wolfram CA consists of lattice sites which can have two possible values, 0 or 1 (*i.e.*, $k = 2$), and which evolve independently of the states of neighbors (*i.e.*, $r = 0$). We can manually construct the four possible rule sets that control the evolution of the CA.

```
{{1} -> 0, {0} -> 0}

{{1} -> 0, {0} -> 1}

{{1} -> 1, {0} -> 0}

{{1} -> 1, {0} -> 1}
```

Looking at one of these rule sets, `{{1} -> 0, {0} -> 1}`, the rule set says that a site whose neighborhood consists of just the site itself, with value 1, is updated to 0 and a site whose neighborhood consists of just the site itself, with value 0, is updated to 1.

In general, each transformation rule in a rule set has a list on the left-hand side and an integer on the right-hand side. The list on the left-hand side of the transformation rule is a tuplet whose elements are values of the $(2r + 1)$ sites in the neighborhood of the site (the middle element of the tuplet is the value of the site). The integer on the right-hand side of the transformation rule is the new value of the site when it is updated during one time step of CA evolution (*e.g.*, the transformation rule `{0, 1, 1} -> 0` says that a site having a value 1 and a left nearest neighbor with value 0 and a right nearest neighbor with value 1, is updated to 0).

As there are, in general, many possible combinations of site values in a neighborhood for given values of k and r, there are many possible rule sets, and the number of rule sets increases very rapidly as k and r increase.

For example, consider the case of $k = 2$ and $r = 1$. This corresponds to having three interacting sites (a given site and its nearest neighbor sites on each side ($v\mathbf{W}x$)), each of which can have two values. The possible neighborhood values for this system are:

```
{{1, 1, 1}, {1, 1, 0}, {1, 0, 1}, {1, 0, 0},

 {0, 1, 1}, {0, 1, 0}, {0, 0, 1}, {0, 0, 0}}
```

Thus, eight or $2^3 = 8$ possible triplets can be formed. Each of these triplets can result in a site value of 1 or 0. For example, one rule set is:

```
{{1,1,1} -> 1, {1,1,0} -> 1, {1,0,1} -> 1, {1,0,0} -> 0,

 {0,1,1} -> 1, {0,1,0} -> 0, {0,0,1} -> 0, {0,0,0} -> 1}
```

Since each triplet can result in either a 1 or 0, and since there are eight possible triplets, there are a total of $2^8 = 256$ rule sets for the $k = 2$, $r = 1$ CA (substantially more than the 4 rule sets for $k = 2$, $r = 0$ CA).

In general, the total number of possible rule sets for a (k, r) Wolfram CA is given by:

```
In[1]:= totalRuleSets[k_, r_] := k^(k^(2r + 1))
```

For the simple cases discussed above ($k = 2$, $r = 0$) and ($k = 2$, $r = 1$), `totalRuleSets` gives:

```
In[2]:= totalRuleSets[2, 0]

Out[2]= 4
```

```
In[3]:= totalRuleSets[2, 1]

Out[3]= 256
```

As two final examples, when each site can have two values and interacts with its two nearest neighbor sites on each side ($uv\mathbf{W}xy$), there are $2^5 = 32$ possible quintuplets and $2^{32} = 4,294,967,296$ rules sets.

```
In[4]:= totalRuleSets[2, 2]

Out[4]= 4294967296
```

When each site can have three values and sites interact with their nearest neighbor site on each side, there are $3^{3^3} = 7,625,597,484,987$ rules sets!

```
In[5]:= totalRuleSets[3, 1]

Out[5]= 7625597484987
```

These computations show that as the number of nearest neighbor sites involved in the interaction increases, the number of rule sets rapidly increases. Furthermore, increasing the number of values that a site can have increases the number of rule sets even more dramatically.

Wolfram has devised a shorthand numerical notation for expressing a rule set for given values of k and r. The numbering scheme can be illustrated using the rule set example given above. The right-hand sides of the transformation rules in this rule set can be used to form the binary number 11101001_2. The base 10 equivalent of this binary number is 233.

```
In[6]:= 2^^11101001
```

```
Out[6]= 233
```

This number is called the *rule number* for the rule set. We will now develop a function which creates a rule set for any rule number and any values of k and r. We start by creating the list of the number of tuplets involved in a (k, r) Wolfram CA.

```
In[7]:= tuplets[k_, r_] :=
            Reverse[Flatten[Apply[Outer,
                Prepend[Table[Range[0, k - 1],
                {(2r + 1)}], List]], 2r]]
```

For example, for the $(2, 1)$ CA, the tuplets function returns:

```
In[8]:= tuplets[2, 1]
```

```
Out[8]= {{1, 1, 1}, {1, 1, 0}, {1, 0, 1}, {1, 0, 0},
          {0, 1, 1}, {0, 1, 0}, {0, 0, 1}, {0, 0, 0}}
```

Next, we create a list, called the *update value* list, of the right-hand sides of the transformation rules in a rule set from a rule number.

The rule number w can be converted into a list whose elements represent the number in base k, using the `IntegerDigits` function. For our example rule set, we have:

```
In[9]:= IntegerDigits[233, 2]
```

```
Out[9]= {1, 1, 1, 0, 1, 0, 0, 1}
```

Using `tuplets[2, 1]` and `IntegerDigits[233, 2]`, we can create the list of transformation rules that constitute the example rule set.

```
In[10]:= MapThread[Rule, {tuplets[2, 1], IntegerDigits[233, 2]}]
```

```
Out[10]= {{1, 1, 1} -> 1, {1, 1, 0} -> 1, {1, 0, 1} -> 1,
           {1, 0, 0} -> 0, {0, 1, 1} -> 1, {0, 1, 0} -> 0,
           {0, 0, 1} -> 0, {0, 0, 0} -> 1}
```

We are now ready to write a general function to convert any rule number w to a rule set for a (k, r) CA, using `tuplets[k, r]` and `IntegerDigits[w, k]`. However, there is a complication. To illustrate the problem, if we use this approach to create the rule set for rule number 24 for the $(2, 1)$ CA, we get an error message.

```
In[11]:= MapThread[Rule,{tuplets[2, 1], IntegerDigits[24, 2]}]


MapThread::mptc:
    Incompatible dimensions of objects at positions
    {2, 1} and {2, 2} of MapThread[Rule, <<1>>];
    dimensions are {8} and {5}.

Out[11]= MapThread[Rule, {{{1, 1, 1}, {1, 1, 0}, {1, 0, 1},
                  {1, 0, 0}, {0, 1, 1}, {0, 1, 0}, {0, 0, 1},
                  {0, 0, 0}}, {1, 1, 0, 0, 0}}]
```

The problem is that lists of different lengths cannot be threaded together, and taking $2^{\wedge\wedge}10000000$ indicates that half of the 256 rule numbers for the (2, 1) CA (those below 128) have less than the prerequisite eight elements needed for threading with the list of the eight triplets for the CA.

We need to add 0s to the front of `IntegerDigits[w, k]` to bring the list length up to k^{2r+1}. This is done as follows:

```
In[12]:= fillCode[k_, r_, w_] :=
              Join[Table[0, {k^(2r + 1) - Length[#]}],
                  #]&[IntegerDigits[w, k]]
```

This will work for all rule numbers. For example,

```
In[13]:= fillCode[2, 1, 233]


Out[13]= {1, 1, 1, 0, 1, 0, 0, 1}


In[14]:= fillCode[2, 1, 24]


Out[14]= {0, 0, 0, 1, 1, 0, 0, 0}
```

Note: The `fillCode` definition takes advantage of the fact that `Table` returns an empty list when the iterator is zero or negative.

Overall then, a rule set can be created from a rule number w, for a (k, r) CA, with the following function:

```
In[15]:= ruleSet[k_, r_, w_] := Module[{fillCode, tuplets},
            fillCode[x_, y_, z_] :=
              Join[Table[0, {x^(2y+1)-Length[#]}],
                #]&[IntegerDigits[z, x]];
            tuplets[p_, s_] :=
              Reverse[Flatten[Apply[Outer,
                    Prepend[Table[Range[0, p-1], {(2s+1)}],
                        List]], 2s]];
            MapThread[Rule,
                {tuplets[k, r], fillCode[k, r, w]}]
          ]
```

For example, using `ruleSet` for the simplest case of a (2, 0) CA, we have

```
In[16]:= {ruleSet[2, 0, 0], ruleSet[2, 0, 1],
          ruleSet[2, 0, 2], ruleSet[2, 0, 3]} //TableForm

Out[16]//TableForm= {1} -> 0    {0} -> 0
                    {1} -> 0    {0} -> 1
                    {1} -> 1    {0} -> 0
                    {1} -> 1    {0} -> 1
```

These are the same four rule sets we created "by hand" earlier.

If `ruleSet[2, 0, 4]` is entered, an error message is returned because there is no rule set corresponding to the number 4 in the case of $k = 2$, $r = 0$.

Now that we have `ruleSet`, we can present the algorithm and code fragments for evolving a one-dimensional cellular automaton.

THE WOLFRAM CA ALGORITHM

We first define the following symbols:

m = number of sites per row

n = maximum number of rows (generations)

w = rule number (this number cannot exceed $k^{k^{2r+1}}$)

k = number of values that a site can assume

$2r + 1$ = number of interacting sites

1. The CA starts either from (a) a row of sites, all of which have value zero except for the center site, which contains a seed (*i.e.*, has a non-zero value),

```
row = ReplacePart[Table[0, {m}],
            Random[Integer, {0, k-1}],
            Ceiling[m/2]]
```

or from (b) a disordered state with randomly distributed seed sites:

```
row = Table[Random[Integer, {0, k-1}], {m}]
```

While the seed sites here have randomly determined values, it is possible to specify the values of the seed sites.

2. The neighborhood around a site in the row is created in two steps:

A. The row of m sites is enlarged to a row of $(m + 2r)$ sites by adding the first r elements of the row to the end of the row, and adding the last r elements of the row to the front of the row.

```
Join[Take[row, -r], row, Take[row, r]]
```

Note: This step is used to implement periodic boundary conditions.

B. The result of step **2.A** is partitioned into m neighborhoods of $(2r + 1)$ overlapping sites, using

```
Partition[Join[Take[row, -r], row,
        Take[row, r]], (2r + 1), 1]
```

3. Applying a rule set to the list of neighborhoods created by step **2.B** gives the updated site values.

```
Partition[Join[Take[row, -r],
            row,
            Take[row, r]], (2r + 1),
        1] /. ruleset
```

4. Finally, the CA evolves using the `FixedPoint` function with an anonymous function to produce new rows of updated sites until there is no change in the values in successive rows or to a maximum of n rows, whichever occurs first.

```
CArows =
    FixedPointList[(Partition[Join[Take[#, -r],
                    #, Take[#, r]], (2r + 1), 1] /.
                    ruleset)&,
                row, n]
```

These code fragments can be put together to create the program.

THE WOLFRAM CA PROGRAM

```
CA[w_, k_, r_, m_, n_] := Module[{ruleSet, row},
    ruleSet = MapThread[Rule,
            {Reverse[Flatten[Apply[Outer,
              Prepend[Table[Range[0, k - 1],
                    {(2r + 1)}], List]], 2r]],
          Join[Table[0, {k^(2r + 1) - Length[#]}],
                #]&[IntegerDigits[w, k]]}];
    row = Table[Random[Integer, {0, k-1}], {m}];
    FixedPointList[(Partition[Join[Take[#, -r], #,
            Take[#, r]], (2r+1),1] /.
        ruleSet)&, row, n]]
```

Rather than using the above program which specifies a disordered initial state and requires that the length of the CA and the number of time steps be specified, we will use a modified version of the program which gives us more flexibility in creating the CA.

The function WolframCA uses options with default values for the length of the initial configuration, the number of time steps, and the initial state. These values can be overridden by giving explicit rules for CASize, CATimeSteps, and InitialState, respectively.

```
In[1]:= Options[WolframCA] = {CASize -> 100, CATimeSteps -> 20,
                            InitialState -> "Disordered"};
```

```
In[2]:= WolframCA[w_, k_, r_, opts___Rule] :=
        Module[{size, steps, state, row, ruleSet},
            size = CASize /. {opts} /. Options[WolframCA];
            steps = CATimeSteps /. {opts} /. Options[WolframCA];
            state = InitialState /. {opts} /. Options[WolframCA];
            row = If[state==="SingleSeed",
                    ReplacePart[Table[0, {size}],
                            k-1, Ceiling[size/2]],
                    Table[Random[Integer, {0, k-1}], {size}]];
            ruleSet = MapThread[Rule,
                            {Reverse[Flatten[Apply[Outer,
                        Prepend[Table[Range[0, k-1], {(2r+1)}],
                                List]], 2r]],
                        Join[Table[0, {k^(2r+1) - Length[#]}],
                            #]&[IntegerDigits[w, k]]}];
            FixedPointList[(Partition[Join[Take[#, -r], #,
                            Take[#, r]], (2r+1), 1] /.
                    ruleSet)&, row, steps]
        ]
```

14.2 ■ RUNNING THE WOLFRAM CA PROGRAM

The function ShowCA is used to graphically represent the cellular automata, where each site value has a different color. Typically, just two values are used for each site ($k = 2$) and the rules depend upon nearest neighbors ($r = 1$).

```
In[3]:= ShowCA[list_List, opts___] :=
         Module[{rule, max = Max[list], colorRule},
            colorRule = Flatten[Map[{# -> Hue[#/(max+1)]}&,
                                       Range[0, max] ]];
            Show[Graphics[
                   RasterArray[Reverse[list] /. colorRule],
                   opts, AspectRatio -> Automatic]]
         ]
```

A few illustrative CAs are shown below. The first four automata start from a random initial configuration and the last graphic shows a CA that starts from a seeded finite state.

This first graphic of rule 136, shows what is typically referred to as "Class I" behavior—changes from the initial configuration occur, but soon die out leading to a fixed, homogeneous state.

```
In[4]:= ShowCA[ WolframCA[136, 2, 1] ]
```

```
Out[4]= -Graphics-
```

The following graphic of rule 4, shows Class II behavior—patterns emerge that consist of periodic regions that are not connected to each other.

In[5]:= `ShowCA[ WolframCA[4, 2, 1] ]`

Out[5]= `-Graphics-`

In the next graphic, showing rule 45, Class III behavior is seen—patterns that are aperiodic and seemingly chaotic are generated. To better see this behavior, we have extended the length of this automaton to 150 sites and run it for 80 iterations.

In[6]:= `ShowCA[ WolframCA[45, 2, 1,`
            `CASize -> 150, CATimeSteps -> 80] ]`

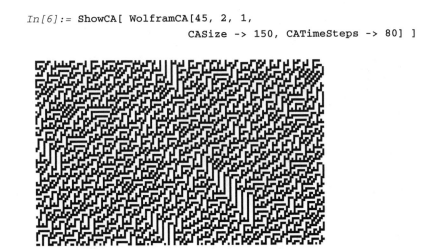

Out[6]= `-Graphics-`

The next graphic shows rule 110, which generates typical Class IV behavior—structures are produced which are complex and quite localized. Again, to better see patterns emerge, we have overridden the default values and explicitly specified an automaton of length 150 running for 60 time steps.

```
In[7]:= ShowCA[ WolframCA[110, 2, 1,
                    CASize -> 150,
                    CATimeSteps -> 60] ]
```

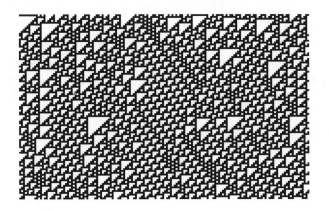

```
Out[7]= -Graphics-
```

Finally, we show an automata that starts from a seeded finite state.

```
In[8]:= ShowCA[WolframCA[30, 2, 1,
                    CASize -> 400,
                    CATimeSteps -> 300,
                    InitialState -> "SingleSeed"]]
```

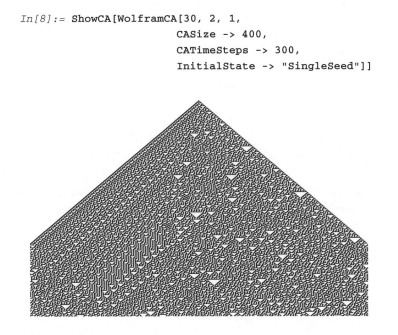

```
Out[8]= -Graphics-
```

ANIMATING THE WOLFRAM CA

While a static figure can be used to detect CA patterns, it is also useful to watch the evolutionary process unfolding.

In creating an animation using `RasterArray` graphics, the main thing that has to be taken into account is that the rectangles, representing site values, that reappear in successive figures in the series that constitute the flip cards of the animation must be the same size and color. This requires two things. First, each figure must have the same number of rows. This is accomplished by adding enough rows of 0s to the `CArows` used in each figure to bring the total number of lines in the figure up to the `Length` of the final CA (so the entire CA computation must be completed before the graphics are created). Second, the transformation rule for converting site values to colors must be applied after all of the site values in all of the CA figures in the animation are calculated and before the graphics command is applied.

Here then, is the CA animation program.

```
In[9]:= AnimateCA[list_List, opts___]:=
            Module[{colorrule, len = Length[list],
                    m = Length[First[list]], k = Max[list]},
                colorrule =
                Flatten[Map[{# -> Hue[#/(k+1)]}&,
                        Range[0,k]]];
                Map[Show[Graphics[RasterArray[#],
                        opts, AspectRatio -> Automatic]]&,
                    Map[Join[Table[Table[0, {m}], {len-#}],
                            Reverse[Take[list,#]]]&,
                        Range[len] ] /. colorrule]
            ]
```

After creating the animation in a *Mathematica* notebook, double-click on the bracket enclosing all of the graphics cells (this bracket is created when **Automatic Grouping** has been selected in the **Cell** menu). This will close up the graphics cells, leaving only the first cell exposed. Double-clicking on the exposed cell starts the animation. The speed of the animation can be adjusted using the number keys on the keyboard (1 is the slowest speed and 9 is the fastest speed).

The following graphic shows the use of `AnimateCA` to illustrate rule 110 for a (2, 1) CA. Thirty time steps are shown on an automaton of length 100.

```
In[10]:= AnimateCA[ WolframCA[110, 2, 1,
                    CASize -> 100,
                    CATimeSteps -> 30] ];
```

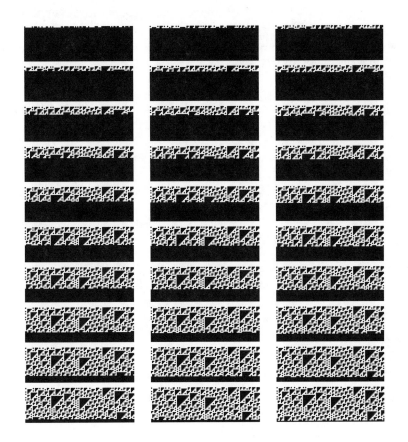

```
Out[10]= -Graphics-
```

14.3 COMPUTER SIMULATION PROJECTS

1. An *attractor* is a pattern or a periodic cycle of patterns to which a set of other patterns evolve. A *basin of attraction* is the set of all the patterns that evolve to an attractor. Experiment with different rule numbers to find an attractor and its basin of attraction.

2. Revise the WolframCA program so that it can be "seeded" with an *arbitrary* initial states. Run a number of CAs starting with your seeded initial configurations.

3. Revise the CA program so that it works without needing to create neighborhoods.

REFERENCES

* S. Wolfram. Computer software in science and mathematics. *Scientific American* 251 (1984) 188–203.

S. Wolfram. Statistical mechanics of cellular automata. *Rev. Mod. Physics* 55 (1983) 601–644.

S. Wolfram. Cellular automata as models of complexity. *Nature* 311 (1984) 419–424.

(Note: These articles are all reprinted in *Cellular Automata and Complexity, Collected Papers* by Stephen Wolfram, Addison-Wesley 1994.

PART

III

Appendices

$\mathcal{A}$PPENDIX A

Mathematica Programming

INTRODUCTION

Mathematica is a programming language of many paradigms. This means that the user can write *Mathematica* programs in a variety of programming styles, from the traditional procedural style to the less conventional (but usually more efficient and elegant) rule-based and functional programming styles. To be able to make an informed choice on which style to adopt for a given computational task, it is necessary to understand how the *Mathematica* language is constructed and how it works. Fortunately, *Mathematica* is fundamentally a simple language that is easily explained and understood. Our purpose here is to present the rudiments of the *Mathematica* programming language in order to enable the reader to read and write *Mathematica* programs.

A.1 EXPRESSIONS

Every quantity entered into *Mathematica* is represented internally as an expression having the form head[arg_1, arg_2, ..., arg_n], where head and arg_i can be other expressions.

For example, if we look at two objects, a list data structure, {a, b, c}, and an arithmetic operation, a + b + c, they appear to be quite different, but if we use the FullForm function to look at their internal representation, we see that they differ only in their heads.

```
In[1]:= FullForm[{a, b, c}]

Out[1]= List[a, b, c]


In[2]:= FullForm[a + b + c]

Out[2]= Plus[a, b, c]
```

The use of a common expression structure to represent everything is not merely cosmetic; it allows us to perform some computations quite simply. For example, to add the elements in a list, it suffices to change the head of the expression, `List`, to `Plus`. This can be done using the built-in `Apply` function, which changes the head of an expression.

```
In[3]:= Apply[Plus, {1, 2, 3}]

Out[3]= 6
```

NONATOMIC EXPRESSIONS

Nonatomic expressions have parts which can be extracted from the expression with the `Part` function, and can be replaced with the `ReplacePart` function.

```
In[4]:= Part[{a, 7, c}, 1]

Out[4]= a


In[5]:= Part[a + b + c, 0]

Out[5]= Plus


In[6]:= ReplacePart[{a, 7, c}, e, 2]

Out[6]= {a, e, c}
```

ATOMIC EXPRESSIONS

Atomic expressions constitute the basic building blocks of the *Mathematica* language. There are three kinds of atomic expressions:

1. A symbol, consisting of a letter followed by letters or numbers (*e.g.*, `darwin`)

2. Four kinds of numbers:

- integer numbers (*e.g.*, 4)
- real numbers (*e.g.*, 5.201)
- complex numbers. (*e.g.*, 3 + 4 I)
- rational numbers (*e.g.*, 5/7)

3. A string, composed of letters, numbers, and spaces (ASCII characters) between quotes (*e.g.*, `"read my lips"`)

Atomic expressions differ from nonatomic expressions in several ways. The `FullForm` of an atomic expression is the atom itself.

In[7]:= `{FullForm[darwin], FullForm["read my lips"], FullForm[5]}`

Out[7]= {darwin, "read my lips", 5}

The Head (or 0th part) of an atom is the type of atom that it is.

In[8]:= `{Head[List], Head["read my lips"], Head[5]}`

Out[8]= {Symbol, String, Integer}

An atomic expression has no parts which can be extracted or replaced.

In[9]:= `Part["read my lips", 1]`

```
Part::partd:
   Part specification read my lips[[1]]
     is longer than depth of object.
```

Out[9]= read my lips[[1]]

COMPOUND EXPRESSIONS

A `CompoundExpression` is a sequence of expressions separated by semicolons.

`expr1; expr2; ...; exprn`

ENTERING AN EXPRESSION

When an expression is entered in *Mathematica,* it is evaluated. The result is returned, unless it is followed by a semicolon.

In[10]:= `2 - 6`

Out[10]= -4

When the entered expression is a compound expression, its components are evaluated sequentially and the result of the last evaluation is returned.

In[11]:= `4 + 5; 7 4`

Out[11]= 28

The details of the steps involved in the evaluation mechanism will be discussed later, but we can easily give you a basic feeling for the evaluation process with the following analogy:

Think of your experiences with using a handbook of mathematical formulas, such as the one shown below.

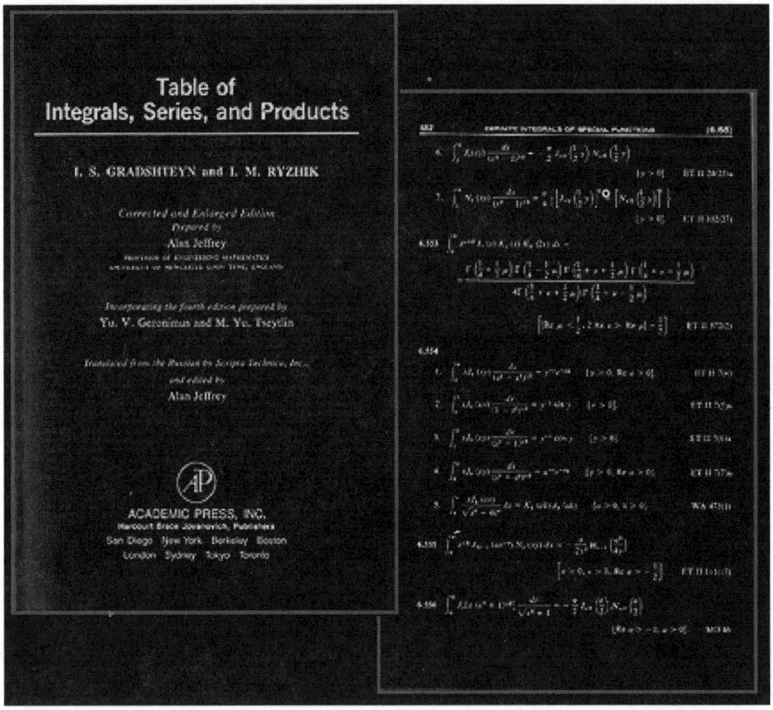

In order to solve an integral, you consult the handbook to locate an integration formula in which the left-hand side has the same form as your integral, except that instead of having specific values for the integrand or the integration limits, it has unspecified (dummy) variables. You then replace your integral with the right-hand side of the formula in the handbook, substituting the specific values from your integral for the corresponding variable symbols. Afterwards, you look through the handbook for formulas (*e.g.*, trigonometric identities or algebraic manipulation) that can be used to change the answer further.

This description makes an excellent analogy to the following description of the *Mathematica* evaluation process.

In *Mathematica*, expression evaluation takes place by term rewriting using rewrite rules. These rules consist of two parts: a pattern on the left-hand side and a replacement text on the right-hand side. When the left-hand side of a rewrite rule is found to pattern-match part of the expression, that part is replaced by the right-hand side of the rule, after substituting values in the expression which match labeled blanks in the pattern into the right-hand side of the rule. Evaluation then proceeds by searching for additional matching rules until no more are found.

If we compare the two descriptions we see that (grammatical differences aside) "formula" is the same as "rewrite rule," "has the same form as" is equivalent to "pattern-matches," "replace" means "term rewriting," and "variable symbol" is "labeled blanks."

The rewrite rules used in evaluation include both the built-in functions and functions that are defined by the user. (Note: You may have noticed that we use the words rule and function interchangeably. This is intentional because, for our purposes here, the two are equivalent.)

Before getting into the details of creating our own user-defined rewrite rules (or programs, as they are more commonly called), we want to first look at the kinds of patterns that *Mathematica* recognizes, since pattern-matching underlies the evaluation process.

A.2 PATTERNS

Patterns are defined syntactically, *i.e.*, by the internal representation of an expression as given using `FullForm`. In general, an expression will be matched by several patterns, of differing specificity. For example, patterns which match x^2, in order of increasing generality, are

1. x raised to the power of 2
2. x raised to the power of a number
3. x raised to the power of something
4. a symbol raised to the power of 2
5. a symbol raised to the power of a number
6. a symbol raised to the power of something
7. something raised to the power of 2
8. something raised to the power of a number
9. something raised to the power of something
10. something

The term "something" used above can be replaced by the term "an expression," so that for example, the last case says that x^2 pattern-matches an expression (which is true, since x^2 is an expression). To be precise, we need a notation to designate a pattern that has the form of an expression . We also need to designate a pattern that has the form of a sequence of expressions, consecutive expressions separated by commas.

Patterns are defined in *Mathematica* as expressions that may contain blanks. A pattern may contain a single (_) blank, a double (__) blank, or a triple (___) blank. These are called the `Blank`, `BlankSequence`, and `BlankNullSequence`, respectively. Their differences will be discussed shortly.

A pattern can be labeled (given a name) so that it can be referred to elsewhere (this is used to make pattern-matching conditional and to represent dummy variables in programs as we will show later). The labeling is done by preceding the blank(s) by a symbol (*e.g.*, `name_`, or `name__`, or `name___`). The labeled pattern is matched by exactly the same expression that matches its unlabeled counterpart and the matching expression is given the name used in the labeled pattern.

Blanks can also be followed by a symbol: `_h`, or `__h`, or `___h`, in which case, an expression must have the head `h` to match the pattern.

PATTERN-MATCHING AN EXPRESSION

We can use the `MatchQ` function to determine whether a particular pattern matches an expression or a sequence of expressions. The most specific pattern-match is between an expression and itself.

```
In[1]:= MatchQ[x^2, x^2]

Out[1]= True
```

To make more general (less specific) pattern-matches, a single blank is used to represent an individual expression. A blank matches any data object. `Blank[h]` or `_h` stands for an expression with the head `h`.

We'll work with x^2 to demonstrate the use of the `Blank` function in pattern-matching. In the following examples (which are arbitrarily chosen from the many possible pattern matches), we'll first state the pattern-match and then check it using `MatchQ`.

x^2 pattern-matches "an expression."

```
In[2]:= MatchQ[x^2, _]

Out[2]= True
```

x^2 pattern-matches "x raised to the power of an expression."

```
In[3]:= MatchQ[x^2, x^_]

Out[3]= True
```

x^2 pattern-matches "x raised to the power of an integer" (or, more formally, "x raised to the power of an expression whose head is `Integer`").

```
In[4]:= MatchQ[x^2, x^_Integer]
Out[4]= True
```

x^2 pattern-matches "an expression whose head is `Power`."

```
In[5]:= MatchQ[x^2, _Power]
Out[5]= True
```

x^2 pattern-matches "an expression whose head is a symbol and which is raised to the power 2."

```
In[6]:= MatchQ[x^2, _Symbol^2]
Out[6]= True
```

x^2 pattern-matches "an expression raised to the power 2."

```
In[7]:= MatchQ[x^2, _^2]
Out[7]= True
```

x^2 pattern-matches "an expression whose head is a symbol and which is raised to the power of an expression whose head is an integer" (or, stated less formally, "a symbol raised to the power of an integer").

```
In[8]:= MatchQ[x^2, _Symbol^_Integer]
Out[8]= True
```

x^2 pattern-matches "an expression raised to the power of an expression."

```
In[9]:= MatchQ[x^2, _^_]
Out[9]= True
```

x^2 pattern-matches "x raised to the power of an expression" (the label y does not affect the pattern-match).

```
In[10]:= MatchQ[x^2, x^y_]
Out[10]= True
```

PATTERN-MATCHING A SEQUENCE

A sequence consists of a number of expressions separated by commas. A double blank represents a sequence of one or more expressions, and __h represents a sequence of one or more expressions, each of which has head h.

For example, a sequence in a list pattern-matches a double blank. (Note: We are pattern-matching the sequence in the list, not the list itself.)

```
In[1]:= MatchQ[{a, b, c}, {__}]

Out[1]= True
```

The arguments of an empty list (which actually has no arguments) do not pattern-match a double blank.

```
In[2]:= MatchQ[{}, {__}]

Out[2]= False
```

An expression that pattern-matches a blank also pattern-matches a double blank. For example,

```
In[3]:= MatchQ[x^2, __]

Out[3]= True
```

A triple blank represents a sequence of zero or more expressions, and ___h represents a sequence of zero or more expressions, each of which has the head h. For example, the triple blank pattern-matches the empty list.

```
In[4]:= MatchQ[{}, {___}]

Out[4]= True
```

An expression that pattern-matches a blank and a sequence that pattern-matches a double blank pattern both pattern-match a triple blank pattern.

```
In[5]:= MatchQ[x^2, ___]

Out[5]= True
```

It is important to be aware that, for the purposes of pattern-matching, a sequence is not an expression. For example,

> *In[6]:=* **MatchQ[{a, b, c}, {_}]**

> *Out[6]=* False

ALTERNATIVE PATTERN-MATCHING

We can make a pattern-match less restrictive by specifying alternative patterns that can be matched.

> *In[1]:=* **MatchQ[x^2, {_} | _^2]**

> *Out[1]=* True

CONDITIONAL PATTERN-MATCHING

We can make a pattern-match more restrictive by making it contingent upon meeting certain conditions. Satisfying these conditions will be a necessary, but not sufficient, requirement for a successful pattern-match.

If the blanks of a pattern are followed by ?test, where test is a predicate (*i.e.*, a function that returns True or False), then a pattern-match is possible only if test returns True when applied to the entire expression. ?test is primarily used with built-in predicate functions.

> *In[1]:=* **MatchQ[2, _Integer?EvenQ]**

> *Out[1]=* True

> *In[2]:=* **MatchQ[2, _Integer?OddQ]**

> *Out[2]=* False

If part of a labeled pattern is followed by /; condition, where condition contains labels appearing in the pattern, then a pattern-match is possible only if condition returns True when applied to the labeled parts of an expression.

> *In[3]:=* **MatchQ[x^2, z_^y_ /; (z=!=y || OddQ[y])]**

> *Out[3]=* True

```
In[4]:= MatchQ[x^2, z_^y_ /; (z=!=y && OddQ[y])]

Out[4]= False
```

With this understanding of the patterns that are recognized by *Mathematica*, we can turn to the details of evaluation.

A.3 EVALUATION

We start by restating the description given earlier of the evaluation process that is used in *Mathematica*.

Evaluation takes place whenever an expression is entered. Evaluation is carried out by term rewriting using rewrite rules. These rules consist of two parts: a pattern on the left-hand side and a replacement text on the right-hand side. When the left-hand side of a rewrite rule is found to pattern-match part of the expression, that part is replaced by the right-hand side of the rule, after values in the expression which match labeled blanks in the pattern are substituted into the right-hand side of the rule. Evaluation then proceeds by searching for further matching rules until no more are found.

The implementation of the evaluation procedure is (with a few exceptions) straightforward:

1. If the expression is a number or a string, it isn't changed.
2. If the expression is a symbol, it is rewritten if there is an applicable rewrite rule in the global rule; otherwise, it is unchanged.
3. If the expression is not a number, string, or symbol, its parts are evaluated in a specific order.

   A. The head of the expression is evaluated.
   B. The arguments of the expression are evaluated from left to right in order. An exception to this occurs when the head is a symbol with a `Hold` attribute (*e.g.*, `HoldFirst`, `HoldRest`, or `HoldAll`), so that some of its arguments are left in their unevaluated forms (unless they, in turn, have the head `Evaluate`). An example of this is the `Set` or `SetDelayed` function, which we will discuss in a moment.

4. After the head and arguments of an expression are each completely evaluated, the expression consisting of the evaluated head and arguments is rewritten, after the arguments are changed, as necessary, based on the `Attributes` (such as `Flat`, `Listable`, `Orderless`) of the head, if there is an applicable rewrite rule in the global rule base.

5. After carrying out the previous steps, the resulting expression is evaluated in the same way and then the result of that evaluation is evaluated, and so on until there are no more applicable rewrite rules.

The details of the term-rewriting process in steps 2 and 4 are as follows:

- part of an expression is pattern-matched by the left-hand side of a rewrite rule.
- the values which match labeled blanks in the pattern are substituted into the right-hand side of the rewrite rule and evaluated.
- the pattern-matched part of the expression is replaced with the evaluated result.

A.4 ■ REWRITE RULES

BUILT-IN FUNCTIONS

Mathematica provides over 1000 built-in functions that can be used for term rewriting. These rules are located in what is referred to as the *global rule base* whenever *Mathematica* is running. Functions defined in a *Mathematica* package are also placed in the global rule base during the session in which the package is loaded. Functions in the global rule base are always available for term rewriting and, importantly, they are always used whenever applicable.

In addition to the built-in rewrite rules, user-defined rewrite rules can be created and placed in the global rule base. These functions are then always available, and always used when applicable, for term rewriting for the duration of the ongoing session. However, they are not automatically preserved beyond the session in which they are created. We want to explain, in detail, how to write user-defined rewrite rules or functions.

USER-DEFINED REWRITE RULES

There are basically two ways to create a user-defined rewrite rule: with the `Set` function and with the `SetDelayed` function.

The `Set` and `SetDelayed` functions have two arguments, the first argument is called the left-hand side (lhs) and the second argument is called the right-hand side (rhs). The `FullForm` representations of these functions are:

```
SetDelayed[lhs, rhs]
```

```
Set[lhs, rhs]
```

When a `SetDelayed` function is entered, it is placed in the global rule base without evaluation of either its left-hand side or right-hand side. On the other hand, when a Set function is entered, its right-hand side is evaluated before being placed in the global rule base. This difference is crucially important and we need to go into it in more detail, so that you will be able to decide which function to use in a given programming situation.

USING THE SET (=) FUNCTION

The Set function is commonly used to make permanent value declarations (or less formally, to give a nickname to a value—a list or a number—which can be used in place of the value). It is written in shorthand notation as

```
lhs = rhs
```

The left-hand side starts with a name, starting with a letter followed by letters and/or numbers with no spaces. The right-hand side is either an expression or a compound expression enclosed in parentheses. The name on the left-hand side may be followed by a set of square brackets containing a sequence of patterns, or labeled patterns, consisting of symbols ending with one or more blanks and the right-hand side may contain the labels on the left-hand side, without blanks. For example, consider the following two simple Set functions:

```
In[1]:= a = {-1, 1}

Out[1]= {-1, 1}


In[2]:= rand1 = Random[Integer, {1, 2}]

Out[2]= 2
```

We can first note that, when a Set function is entered, a value is returned, unless it is followed by a semicolon. We can look into the global rule base to see what rewrite rules have been created when a and rand1 were entered.

```
In[3]:= ?a

Global`a
a = {-1, 1}


In[4]:= ?rand1

Global`rand1
rand1 = 2
```

We find that the rewrite rule associated with a is the same as the `Set` function we entered, but the rewrite rule associated with `rand1` differs from the corresponding `Set` function. The reason for this is that when a `Set` function is entered, its left-hand side is left unevaluated while its right-hand side is evaluated. This property is known as the `HoldFirst` attribute.

```
In[5]:= Attributes[Set]

Out[5]= {HoldFirst, Protected}
```

```
In[6]:= ?HoldFirst

HoldFirst is an attribute which specifies that
    the first argument to a function is to be
    maintained in an unevaluated form.
```

When a rewrite rule is created from a `Set` function, the unevaluated left-hand side and the evaluated right-hand side of the function are used.

In the above cases, the evaluation of the right-hand side of a resulted in the list {-1, 1} and the evaluation of the right-hand side of `rand1` resulted in 2.

When the right-hand side is a compound expression enclosed in parentheses, the expressions of the right-hand side are evaluated in sequence and the right-hand side of the resulting rewrite rule is the result of the final evaluation.

```
In[7]:= rand2 = (b = {-1, 1}; Random[Real, b])

Out[7]= -0.230583
```

```
In[8]:= ?rand2

Global`rand2
rand2 = -0.2305826544105006707
```

What happened here is that b was first evaluated to give {-1, 1} and this value was then used to evaluate the random number function.

Note that in this simple example, the `Set` function associated with `rand2` contains another `Set` function, associated with b, within it. If we query the rule base for information about b, we find that a rewrite rule associated with b has been created.

```
In[9]:= ?b

Global`b
b = {-1, 1}
```

In general, when a Set function is entered, both it and any Set or SetDelayed functions on the right-hand side create rewrite rules in the global rule base.

After a value has been declared by entering a Set function, the appearance of the value's name during an evaluation causes the value itself to be substituted in (which is why we say that it acts like a nickname). The rewrite rule associated with rand2 in the global rule base is used below as an argument to the Abs function.

```
In[10]:= Abs[rand2]

Out[10]= 0.230583
```

The left-hand side of a rewrite rule can be associated with only one value at a time. When a Set function is entered, the resulting rewrite rule overwrites any previous rewrite rule with the identical left-hand side. For example,

```
In[11]:= rand3 = {-1, 1}[[Random[Integer, {1, 2}]]];


In[12]:= ?rand3

Global'rand3
rand3 = -1


In[13]:= rand3 = {-1, 1}[[Random[Integer, {1, 2}]]];


In[14]:= ?rand3

Global'rand3
rand3 = 1
```

What we see is that the value of rand3 was -1 after rand3 was first entered and this value was then changed to 1 after rand3 was reentered.

Here is a simple compound expression.

```
In[15]:= (rand4 = {-1,1}[[Random[Integer, {1,2}]]]; Print[rand4];
          rand4 = rand4 + {-1,1}[[Random[Integer, {1,2}]]])

Out[15]= -1
Out[15]= -2


In[16]:= ?rand4

Global'rand4
rand4 = -2
```

The `Print` statement above allows us to see the result of the evaluation of an intermediate expression in a compound expression (which is normally not shown). What we see is that `rand4` was first evaluated to −1 and this value was used in the subsequent reevaluation of `rand4`, which yielded −2, and this value was placed in the global rule base, replacing the previous value of −1.

While the left-hand side of a rewrite rule can be associated with only one value at a time, a value can be associated with several names simultaneously. We made use of this earlier when we defined both a and b as {-1, 1}. As we can see, both names are entered into the global rule base with the same value.

```
In[17]:= ?a

Global'a
a = {-1, 1}


In[18]:= ?b

Global'b
b = {-1, 1}
```

Finally, user-defined rewrite rules can be removed from the global rule base using either the `Clear` or `Remove` function.

```
In[19]:= Clear[b]


In[20]:= ?b

Global'b
```

USING THE SETDELAYED (:=) FUNCTION

The `SetDelayed` function is commonly used in writing function definitions (programs). It is written in shorthand notation as

```
left-hand side := right-hand side
```

The left-hand side starts with a name. The name is usually (but not necessarily) followed by a set of square brackets containing a sequence of patterns, which are usually labeled, consisting of symbols ending with one or more blanks. The right-hand side is either an expression or a compound expression enclosed in parentheses, which may contain the names on the left-hand side, without blanks. We'll use the

`SetDelayed` function in the form it usually takes when a *Mathematica* program is written:

```
name[arg1_, arg2_, ..., argn_] :=
                (expr1; expr2; ... ; exprm)
```

where the arguments, *argi*, are pattern labels (dummy variables) that may appear in the compound expression on the right-hand side. Any number of variables can be used in the sequence.

For example, consider the function definition

```
In[1]:= f[x_] := x^2
```

The first thing we notice is that, in contrast to a `Set` function, nothing is returned when a `SetDelayed` function is entered. If we query the rule base, we see that a rewrite rule associated with f has been placed in the global rule base that is identical to the `SetDelayed` function.

```
In[2]:= ?f

Global`f
f[x_] := x^2
```

The reason is that when a `SetDelayed` function is entered, both its left-hand side and the right-hand side are left unevaluated. This property is known as the `HoldAll` attribute.

```
In[3]:= Attributes[SetDelayed]

Out[3]= {HoldAll, Protected}
```

```
In[4]:= ?HoldAll

HoldAll is an attribute which specifies that all arguments
    to a function are to be maintained in an unevaluated form.
```

Each time the left-hand side of a `SetDelayed` rewrite rule is entered with specific argument values, the right-hand side of the rule is evaluated using these values, and the result is returned. For example,

```
In[5]:= f[8]

Out[5]= 64
```

Note that the right-hand side is evaluated anew each time the left-hand side is entered with specific argument values. Thus, in our example, entering f[3] yields 9.

```
In[6]:= f[3]

Out[6]= 9
```

The user-defined function, f, is called in the same way as a built-in function, by entering its name with specific argument value(s).

When the right-hand side of the SetDelayed function is a compound expression enclosed in parentheses, no rewrite rules are created from the auxiliary functions on the right-hand side when the function is entered, because the right-hand side is not evaluated. When the program is run (or equivalently, the user-defined function is called) for the first time, all of the auxiliary functions are then placed in the global rule base.

PLACING CONSTRAINTS ON A REWRITE RULE

The use of a rewrite rule can be restricted by attaching constraints on either the left-hand side or the right-hand side of a SetDelayed rule. Conditional pattern-matching with _h or with _? and _/; can be attached to the dummy variable arguments on the left-hand side. Also, /; can be placed on the right-hand side, immediately after the (compound) expression.

LOCALIZING NAMES IN A REWRITE RULE

As we have pointed out, when the right-hand side of a Set or SetDelayed function is evaluated (which occurs when a Set function is first entered and when a SetDelayed rewrite rule is first called), rewrite rules for all of its auxiliary functions are placed in the global rule base. This can cause a problem if a name being used in a program conflicts with the use of the name elsewhere.

We can prevent a name clash by insulating the auxiliary functions within the rewrite rule so that they are not placed in the global rule base as separate rewrite rules and they will then exist only while being used in the evaluation of the rule. This is usually done using the Module function.

```
left-hand side :=
    Module[{name1 = val1, name2, ...}, right-hand side]
```

The Module form is the accepted way of writing a *Mathematica* program containing auxiliary function definitions and value declarations.

CREATING REWRITE RULES DYNAMICALLY

One way to speed up the evaluation of an expression in *Mathematica* is to have it remember the values it computes. This can be done by creating `Set` rewrite rules during the evaluation of a `SetDelayed` rewrite rule. To do this, a `SetDelayed` function is written in which the right-hand side is a `Set` function of the same name. The general form is:

```
f[arg_] := f[arg] = right-hand side
```

As an example of this "dynamic programming," consider the definition and computation of the Fibonacci numbers.

```
In[1]:= fib[0] := 0
        fib[1] := 1
        fib[n_] := fib[n] = fib[n-1] + fib[n-2]
```

When these rewrite rules are entered, the three rewrite rules are placed in the global rule base.

```
In[2]:= ?fib
Global`fib
fib[0] := 0
fib[1] := 1
fib[n_] := fib[n] = fib[n - 1] + fib[n - 2]
```

When the program is subsequently run for n = 3 say, the final value of `fib[3]` and the intermediate value of `fib[2]` are both computed and, as these values are determined, they are entered as rewrite rules into the global rule base, as we can see by querying the global rule base.

```
In[3]:= fib[3]

Out[3]= 2
```

```
In[4]:= ?fib
Global`fib
fib[0] := 0
fib[1] := 1
fib[2] = 1
fib[3] = 2
fib[n_] := fib[n] = fib[n - 1] + fib[n - 2]
```

The main benefit of this method is that having these additional rules in the global rule base saves future computing time since the new rewrite rules can be used when applicable, obviating the need to recalculate previously determined values. The use of dynamic programming is especially useful for recursive programs.

ORDERING REWRITE RULES

When the left-hand side of more than one built-in or user-defined rewrite rule is found to pattern-match an expression, the choice of which rule to use is determined by the order of precedence:

- A user-defined rule is used before a built-in rule.
- A more specific rule is used before a more general rule (the more specific the rule, the fewer expressions it pattern-matches).

So, for example, if we have two rewrite rules whose left-hand sides have the same name but whose labeled patterns have different specificity, both rules will appear in the global rule base (since the left-hand sides are not identical) and the more specific rule will be used in preference to the more general rule. For example, here are two rules for a function f:

```
In[1]:= f[x_] := x^2
```

```
In[2]:= f[x_Integer] := x^3
```

We now query the rule base:

```
In[3]:= ?f
Global`f
f[x_Integer] := x^3
f[x_] := x^2
```

Now, entering f with a real-valued argument returns a different result from entering f with an integer-valued argument.

```
In[4]:= f[6.]
Out[4]= 36.
```

```
In[5]:= f[6]
Out[5]= 216
```

While a specific real-valued argument pattern-matches both `x_` and `x_Integer` (and hence pattern-matches both of the `f` rewrite rules), the latter rule is a more specific pattern-match for the integer value 6.

If *Mathematica* cannot decide which rule is more general, it uses the rules in the order in which they were entered into the global rule base.

The ordering of rewrite rules makes it possible for us to create sets of rewrite rules with the same name that give different results, depending on the arguments used. This is key to writing rule-based programs.

ASSOCIATING REWRITE RULES WITH SYMBOLS

When a rewrite rule is placed in the global rule base, it is associated with (or attached to) the leftmost symbol on the left-hand side of the rewrite rule (the definition is said to be a *downvalue* of the associated symbol). For example,

```
In[1]:= s[t[y_]] := right-hand side[y]
```

```
In[2]:= ?s
```

```
Global`s
s[t[y_]] := right-hand side[y]
```

```
In[3]:= ?t
```

```
Global`t
```

This is why, when you enter an expression such as the following, an error message is returned indicating that you are improperly attempting to associate the rewrite rule with the protected symbol `Plus`.

```
In[4]:= f[x_] + g[ x_] := f[g[x]]
```

```
SetDelayed::write: Tag Plus in f[x_] + g[x_] is Protected.
```

```
Out[4]= $Failed
```

To associate a rule with a symbol other than the leftmost symbol, the `Upvalue` function can be used (the definition is said to be an *upvalue* of the associated symbol). For example,

```
In[5]:= u/: v[u[y_]] := right-hand side[y]
```

```
In[6]:= ?v

Global`v

In[7]:= ?u

Global`u
v[u[y_]] ^:= right-hand side[y]
```

When there is a choice as to which symbol to associate with a rewrite rule, it is preferable to use the less-common symbol.

A.5   TRANSFORMATION RULES

There are times when we want a rewrite rule to be applied to (used inside) only one specific expression, rather than being placed in the global rule base where it will be used whenever it pattern-matches an expression. For example, the temporary substitution of a value for a name in an expression may be preferable to the permanent assignment of the name to the value via a Set function. When this is the case, the ReplaceAll function can be used together with a Rule or RuleDelayed function to create a transformation (or local rewrite) rule which is placed directly after the expression to which it is to be applied. (Note: It can be shown that any procedural program can also be done using transformation rules.)

In deciding whether to use the Rule or RuleDelayed function to create a transformation rule, it is necessary to understand how each function is evaluated because, while both functions are applied to an expression after the expression is evaluated, the functions themselves are evaluated differently.

USING THE RULE (–>) FUNCTION
The Rule function has the general form

```
Rule[lhs, rhs]
```

The shorthand notation for writing an expression with its attached transformation rule(s) is:

```
expression /. lhs -> rhs
```

or

```
expression /. {lhs1 -> rhs1, lhs2 -> rhs2, ...}
```

The left-hand side of a `Rule` transformation rule can be written using symbols, numbers, or labeled patterns. Both the left-hand side and right-hand side of a `Rule` transformation rule are evaluated and the results are used in the evaluated expression. For example, here is a list:

```
In[1]:= Table[x, {4}]/. x -> (-1)^Random[Integer]

Out[1]= {1, 1, 1, 1}
```

This `Trace` function traces the evaluation process:

```
In[2]:= Trace[Table[x, {4}]/. x -> (-1)^Random[Integer]]

Out[2]= {{Table[x, {4}], {x, x, x, x}},
                                            0
            {{{Random[Integer], 0}, (-1) , 1}, x -> 1, x -> 1},
         {x, x, x, x} /. x -> 1, {1, 1, 1, 1}}}
```

We can see that `(-1)^Random[Integer]` was evaluated before it was substituted for `x` in the list and hence all of the elements in the list are identical.

There are two important things to remember when using multiple transformation rules with an expression. When a list of transformation rules appears on the right-hand side of `/.` the rules are applied in order so that a later rule in the list is used only if none of the previous rules match. Furthermore, only one transformation rule, at most, is applied to a given part of an expression, and thereafter no matching rules will be used.

USING THE RULEDELAYED (:>) FUNCTION

The `RuleDelayed` function has the general form

```
Rule[lhs, rhs]
```

The shorthand notation for writing an expression with its attached transformation rule is:

```
expression /. lhs :> rhs
```

The left-hand side of a `RuleDelayed` transformation rule can be written using symbols, numbers, or labeled patterns. It is evaluated before it is used in the evaluated expression, but the right-hand side is not evaluated until after it is substituted into the evaluated expression. For example, here is a list:

```
In[1]:= Table[x, {4}]/. x :> (-1)^Random[Integer]

Out[1]= {-1, 1, 1, -1}
```

Here is the `Trace` of the evaluation:

```
In[2]:= Trace[Table[x, {4}]/. x :> (-1)^Random[Integer]]

Out[2]= {{Table[x, {4}], {x, x, x, x}},
                                        Random[Integer]
            {x, x, x, x} /. x :> (-1)                   ,
                Random[Integer]         Random[Integer]
            {(-1)                , (-1)                ,
              Random[Integer]         Random[Integer]
            (-1)                , (-1)                },
                                    1
            {{Random[Integer], 1}, (-1) , -1},
                                    0
            {{Random[Integer], 0}, (-1) , 1},
                                    0
            {{Random[Integer], 0}, (-1) , 1},
                                    1
            {{Random[Integer], 1}, (-1) , -1}, {-1, 1, 1, -1}}
```

The unevaluated expression, `(-1)^Random[Integer]`, was first substituted for `x` in the list and then each occurrence of the expression in the list was evaluated, resulting in a list whose elements have varying values.

PLACING CONSTRAINTS ON A TRANSFORMATION RULE

By placing `/; condition` immediately after a `RuleDelayed :>` transformation rule, its use can be restricted in the same way that using `/; condition` can be used to restrict the use of a `SetDelayed` rewrite rule.

APPLYING A TRANSFORMATION RULE REPEATEDLY

A transformation rule is applied only once to each part of an expression (in contrast to a rewrite rule) and multiple transformation rules are used in parallel. For example,

```
In[1]:= {a, b, c}/.{c -> b, b -> a}

Out[1]= {a, a, b}
```

The `ReplaceRepeated` function is used to apply one or more transformation rules repeatedly to an expression until the expression no longer changes. For example,

```
In[2]:= {a, b, c}//.{c -> b, b -> a}

Out[2]= {a, a, a}
```

In using //. with a list of transformation rules, it is important to keep in mind the order of application of the rules. The transformation rules are not repeatedly applied in order; rather, each rule, in turn, is applied repeatedly.

CONTROLLING THE EVALUATION ORDER

Our discussion of the rewrite and transformation rules has emphasized the importance of understanding the order of evaluation used by *Mathematica*. It is therefore appropriate to end this section by pointing out that the user can (to some extent) wrest control of the evaluation process from *Mathematica* and either force or prevent evaluation. We won't go into the details of doing this, but we can indicate functions that can be used for this purpose: Hold, HoldAll, HoldFirst, HoldRest, HoldForm, HeldPart, ReleaseHold, Evaluate, Unevaluated, and Literal.

A.6 ■ FUNCTIONS

Mathematica works with functions, both built-in and user-defined, in ways that make them very powerful programming tools. These capabilities, which are characteristic of the functional style of programming, include:

- User-defined functions can be created and used "on the spot" without being named or entered before being used. (These are called *anonymous functions*.)
- There are built-in functions which can take functions as arguments and/or return functions as results. (These are known as *higher-order* functions.)
- Built-in functions are automatically applied in parallel to the elements of their list argument(s) and user-defined functions can also be be given this property. (This is known as the Listable attribute.)
- Functions can be applied successively without having to declare the value of the result of one function call before using it as an argument in another function call. (This is known as making a *nested function call*.)

We can give some simple illustrations of these functional features.

ANONYMOUS FUNCTIONS

An *anonymous* function is used to represent a function without having to give it a name. An anonymous function can be applied at the moment it is created without having first to be entered as a definition. The general form of an anonymous function with one argument is:

```
Function[x, body]
```

When there is more than one argument, the form is:

```
Function[{x, y, ...}, body]
```

There is simpler notation that we often use. An anonymous function is written using the same form as the right-hand side of a rewrite rule, replacing variable symbols with #1, #2, ..., and enclosing the expression in parentheses followed by an ampersand.

This notation can be demonstrated by converting some simple user-defined functions into anonymous functions. For example, a rewrite rule that squares a value

```
In[1]:= square[x_] := x^2
```

can be written as

```
(#^2)&
```

and it can be applied to an argument, *e.g.*, 5, instantly:

```
In[2]:= (#^2)&[5]
Out[2]= 25
```

The following example of an anonymous function with two arguments, raises the first argument to the power of the second argument.

```
(#1^#2)&
```

```
In[3]:= (#1^#2)&[5, 3]
Out[3]= 125
```

It is important to distinguish between an anonymous function which takes multiple arguments and an anonymous function which takes a list with multiple elements as its argument.

For example, the anonymous function just given doesn't work with an ordered pair argument:

```
In[4]:= (#1^#2)&[{2, 3}]
```

```
Out[4]= Function::slotn:
                                #2
           Slot number 2 in #1    &

           cannot be filled from

                      #2
           (#1    & )[{2, 3}].
```

If we want to perform the operation on the components of an ordered pair, the appropriate anonymous function is:

```
In[5]:= (#[[1]]^#[[2]])&[{2, 3}]
```

```
Out[5]= 8
```

Anonymous functions are particularly valuable when used with higher-order functions.

HIGHER-ORDER FUNCTIONS

A higher-order function takes a function as an argument and/or returns a function as a result. This is known as "treating functions as first-class objects." We'll illustrate the use of some of the most important built-in higher-order functions.

APPLY

```
In[1]:= ?Apply
```

```
Apply[f, expr] or f @@ expr replaces the head of expr by f.
Apply[f, expr, levelspec] replaces heads in parts of
expr specified by levelspec.
```

We have already seen `Apply` used to add the elements of a linear list. Given a nested list argument, `Apply` can be used on the outer list or the interior lists. Here is an example with a general function f, and a nested list.

```
In[2]:= Apply[f, {{a, b}, {c, d}}]
```

```
Out[2]= f[{a, b}, {c, d}]
```

```
In[3]:= Apply[f, {{a, b}, {c, d}}, 2]
```

```
Out[3]= {f[a, b], f[c, d]}
```

MAP

In[1]:= **?Map**

Map[f, expr] or f /@ expr applies f to each element on the
first level in expr.

Map[f, expr, levelspec] applies f to parts of expr
specified by levelspec.

Here is an example with a general function f, and a linear list.

In[2]:= **Map[f, {a, b, c, d}]**

Out[2]= {f[a], f[b], f[c], f[d]}

For a nested list structure, Map can be applied to either the outer list or to the
interior lists, or to both.

In[3]:= **Map[g, {{a, b}, {c, d}, {e, f}}]**

Out[3]= {g[{a, b}], g[{c, d}], g[{e, f}]}

In[4]:= **Map[g, {{a, b}, {c, d}, {e, f}}, {2}]**

Out[4]= {{g[a], g[b]}, {g[c], g[d]}, {g[e], g[f]}}

In[5]:= **Map[g, {{a, b}, {c, d}, {e, f}}, 2]**

Out[5]= {g[{g[a], g[b]}], g[{g[c], g[d]}], g[{g[e], g[f]}]}

Note that the Map function does not return {g[a, b], g[c, d], g[e,
f]} in any of the above examples. To obtain that result, we can use Map with an
anonymous function containing Apply.

In[6]:= **Map[Apply[g, #]&,{{a, b}, {c, d}, {e, f}}]**

Out[6]= {g[a, b], g[c, d], g[e, f]}

MAPTHREAD

```
In[1]:= ?MapThread
```

MapThread[f, {{a1, a2, ...}, {b1, b2, ...}, ...}] gives {f[a1, b1, ...], f[a2, b2, ...], ...}.

MapThread[f, {xa, xb, ...}, n] maps f over the nth level of the n-dimensional tensors xa, xb, ...

Here is an example with a general function g, and a nested list.

```
In[2]:= MapThread[g, {{a, b, c}, {x, y, z}}]
```

```
Out[2]= {g[a, x], g[b, y], g[c, z]}
```

One useful example of this function is the pairing-off of the corresponding elements of lists into ordered pairs (which is done faster this way than by using the Transpose function).

```
In[3]:= MapThread[List,{{a, b ,c }, {d, e, f}}]
```

```
Out[3]= {{a, d}, {b, e}, {c, f}}
```

LISTABILITY

The built-in functions have the property known as the Listable attribute. For example,

```
In[1]:= Log[{a, b, c, d}]
```

```
Out[1]= {Log[a], Log[b], Log[c], Log[d]}
```

```
In[2]:= Plus[{a, b}, {c, d}]
```

```
Out[2]= {a + c, b + d}
```

We can also make a user-defined function Listable. For example, the general function, h, can be made Listable.

```
In[3]:= Attributes[h] = Listable;
```

We can demonstrate how the `Listable` attribute works by performing operations using h and then showing how the same results can be obtained with a general unlistable function hh.

```
In[4]:= h[{a, b, c, d}]
Out[4]= {h[a], h[b], h[c], h[d]}
```

```
In[5]:= Map[hh, {a, b, c, d}]
Out[5]= {hh[a], hh[b], hh[c], hh[d]}
```

```
In[6]:= h[{{a, b}, {c, d}}]
Out[6]= {{h[a], h[b]}, {h[c], h[d]}}
```

```
In[7]:= Map[hh, {{a, b}, {c, d}}, {2}]
Out[7]= {{hh[a], hh[b]}, {hh[c], hh[d]}}
```

```
In[8]:= h[{a, b}, {c, d}]
Out[8]= {h[a, c], h[b, d]}
```

```
In[9]:= Thread[hh[{a, b}, {c, d}]]
Out[9]= {hh[a, c], hh[b, d]}
```

It is also possible to perform this last operation by putting the list arguments inside another list and using the `MapThread` function.

```
In[10]:= MapThread[hh, {{a, b}, {c, d}}]
Out[10]= {hh[a, c], hh[b, d]}
```

FOLDLIST AND FOLD

```
In[1]:= ?FoldList
FoldList[f, x, {a, b, ...}] gives {x, f[x, a], f[f[x, a], b], ...}.
```

The `Fold` operation takes a function, a value, and a list, applies the function to the value, and then applies the function to the result and the first element of the list, and then applies the function to the result and the second element of the list and so on.

In[2]:= **Fold[Plus, 0, {a, b, c ,d}]**

Out[2]= a + b + c + d

In[3]:= **FoldList[Plus, 0, {a, b, c ,d}]**

Out[3]= {0, a, a + b, a + b + c, a + b + c + d}

NestList and Nest

In[1]:= **?NestList**

NestList[f, expr, n] gives a list of the results of
applying f to expr 0 through n times.

The Nest operation applies a function to a value, then applies the function to the result, and then applies the function to that result, and so on, a specified number of times.

In[2]:= **NestList[g, a, 4]**

Out[2]= {a, g[a], g[g[a]], g[g[g[a]]], g[g[g[g[a]]]]}

In[3]:= **NestList[Sin, 0.7, 3]**

Out[3]= {0.7, 0.644218, 0.600573, 0.565115}

which is equivalent to

In[4]:= **{0.7, Sin[0.7], Sin[Sin[0.7]], Sin[Sin[Sin[0.7]]]}**

Out[4]= {0.7, 0.644218, 0.600573, 0.565115}

If we are only interested in the final result of the NestList operation, we can use the Nest function, which does not return the intermediate results.

In[5]:= **Nest[g, a, 4]**

Out[5]= g[g[g[g[a]]]]

FixedPointList and FixedPoint

The Nest operation does not stop until it has completed a specified number of function applications. There is another function which performs the Nest operation, stopping after whichever of the following occurs first: (a) there have been a specified number of function applications, (b) the result stops changing, or (c) some predicate condition is met.

```
In[1]:= ?FixedPointList
```

```
FixedPointList[f, expr] generates a list giving the results
of applying f repeatedly, starting with expr, until the
results no longer change.
FixedPointList[f, expr, n] stops after at most n steps.
```

Here is an example.

```
In[2]:= FixedPointList[Sin, 0.7, 3,
                        SameTest -> (#2 < 0.65 &)]
```

```
Out[2]= {0.7, 0.644218}
```

```
In[3]:= FixedPointList[Sin, 0.7, 3,
                        SameTest -> ((#1 - #2) < 0.045 &)]
```

```
Out[3]= {0.7, 0.644218, 0.600573}
```

In these examples, #1 refers to the next-to-last element in the generated list and #2 refers to the last element in the list.

Nested Function Calls

A nested function call is an application of a function to the result of applying another function to some argument value.

Consider the following consecutive computations:

```
In[1]:= Tan[4.0]
```

```
Out[1]= 1.15782
```

```
In[2]:= Sin[%]
```

```
Out[2]= 0.915931
```

```
In[3]:= Cos[%]
```

```
Out[3]= 0.609053
```

We can combine these function calls into a nested function call

```
In[4]:= Cos[Sin[Tan[4.0]]]

Out[4]= 0.609053
```

Notice that the result of one function call is immediately fed into another function without the need to first name (or declare) the result.

Nested function calls are a generalization of `NestList`, `FoldList`, and `FixedPointList`, which repeatedly apply a single function.

As an example, a nested function call is used to create a deck of playing cards:

```
In[5]:= cardDeck = Flatten[Outer[List, {c,d,h,s},
                    Join[Range[2,10], {J,Q,K,A}]],
                1]

Out[5]= {{c, 2}, {c, 3}, {c, 4}, {c, 5}, {c, 6},
          {c, 7}, {c, 8}, {c, 9}, {c, 10},
          {c, J}, {c, Q}, {c, K}, {c, A}, {d, 2},
          {d, 3}, {d, 4}, {d, 5}, {d, 6}, {d, 7},
          {d, 8}, {d, 9}, {d, 10}, {d, J},
          {d, Q}, {d, K}, {d, A}, {h, 2}, {h, 3},
          {h, 4}, {h, 5}, {h, 6}, {h, 7}, {h, 8},
          {h, 9}, {h, 10}, {h, J}, {h, Q},
          {h, K}, {h, A}, {s, 2}, {s, 3}, {s, 4},
          {s, 5}, {s, 6}, {s, 7}, {s, 8}, {s, 9},
          {s, 10}, {s, J}, {s, Q}, {s, K}, {s, A}}
```

NESTING ANONYMOUS FUNCTIONS

Anonymous functions can be nested, in which case it is sometimes necessary to use the `Function[var, body]` form rather than the `(... # ...)&` form, in order to distinguish between the arguments of the different anonymous functions.

```
In[1]:= (Map[(#^2)&, #])&[{3, 2, 7}]

Out[1]= {9, 4, 49}


In[2]:= Function[y, Map[Function[x, x^2], y]][{3, 2, 7}]

Out[2]= {9, 4, 49}
```

The two forms can also be used together.

```
In[3]:= Function[y, Map[(#^2)&, y]][{3, 2, 7}]

Out[3]= {9, 4, 49}


In[4]:= (Map[Function[x, x^2], #])&[{3, 2, 7}]

Out[4]= {9, 4, 49}
```

REFERENCES

R. J. Gaylord, S. N. Kamin, and P. R. Wellin. *Introduction to Programming with Mathematica.* TELOS/Springer-Verlag NY 1993.

R. Maeder. *Programming in Mathematica*, Second Edition. Addison-Wesley 1991.

S. Wolfram. *Mathematica: A System for Doing Mathematics by Computer*, Second Edition. Addison-Wesley 1992.

*A*PPENDIX **B**

Random Numbers

Probability is the most important concept in modern science, especially as nobody has the slightest notion what it means.

— Bertrand Russell

WHAT IS A RANDOM NUMBER?

We often think of choosing a random number as selecting one which is as likely to occur as any other. Although each of us probably has a good intuitive sense of what we mean by *random number*, the formal, mathematical notion can be a bit slippery. In fact, there really is no such thing as "a single random number"—we usually think of *sequences of random numbers*. In this sense, any number in the sequence should have nothing to do with any of the other numbers in the sequence. So for example, the sequence of numbers 1, 1, 2, 3, 5, 8, ..., is clearly not a random sequence in the above sense, as each number is intimately tied to others in the sequence (each number is the sum of its two predecessors). On the other hand, the following sequence of integers appears to be random in this informal sense:

{59, 68, 65, 32, 100, 25, 56, 84, 82, 49, 18, 38, 14,
   59, 61, 65, 79, 91, 7, 43, 49, 63, 58, 70, 77}

Computers generate sequences of random numbers according to some pre-scribed algorithm. By design, such sequences are purely deterministic. Use the same algorithm, start with the same initial configuration, and you will get the exact same sequence. For this reason, it is common to call such numbers *pseudorandom* numbers; for our purposes though, we shall be a little loose with the terminology and call them random.

All computer languages use random number generators to produce sequences in which each number is obtained from its successor by repeated application of a well-defined algorithm. These sequences start with a *seed* and apply the algorithm repeatedly to generate numbers.

One of the more well-known random number generators is the *linear congruential method*. Starting with a seed x_0, successive numbers in the sequence are determined by

$$x_n = (ax_{n-1} + c) \pmod{m}$$

where a is referred to as the *multiplier*, c is called the *increment*, and m is called the *modulus*. Although the linear congruential method can generate fairly good sequences of random numbers, it can also fall into some short cycles. For example,

```
In[1]:= LinearCongruential[n_] :=
            Mod[a LinearCongruential[n-1] + c, m]
```

```
In[2]:= a = 3;
        c = 6;
        m = 64;
        LinearCongruential[0] = 1;
        Map[LinearCongruential, Range[0, 12]]
```

```
Out[2]= {1, 9, 33, 41, 1, 9, 33, 41, 1, 9, 33}
```

Choosing a and c relatively prime (having no factors in common) will, in general, give longer periods:

```
In[3]:= a = 5;
        c = 6;
        m = 64;
        LinearCongruential[0] = 1;
        Map[LinearCongruential, Range[0, 40]]
```

```
Out[3]= {1, 11, 61, 55, 25, 3, 21, 47, 49, 59, 45,
         39, 9, 51, 5, 31, 33, 43, 29, 23, 57, 35,
         53, 15, 17, 27, 13, 7, 41, 19, 37, 63, 1,
         11, 61, 55, 25, 3, 21, 47, 49}
```

What is easy to see is that a small modulus can lead to short cycles. Even a relatively large modulus can lead to short cycles if you are not careful in how the increment and multiplier are chosen. (For a full treatment of the linear congruential method, see Knuth 1981.) It is also quite easy to see that this method is completely deterministic. Start with the same seed, and the same sequence will be generated.

Although you can be much more careful in the choice of parameters (in particular, choosing a large modulus), and you can try variants such as a quadratic congruential method (where $x_n = ax_{n-1}^2 + bx_{n-1} + c \pmod{m}$), the main point to keep in mind is that such methods are decidedly deterministic.

B.1 ■ GENERATING RANDOM NUMBERS WITH *MATHEMATICA*

Mathematica can be used to create random numbers with arbitrary probability distributions. The numbers can be integer, real, or complex. There are dozens of different distributions that you can use, or you can create your own. By default, *Mathematica* uses a uniform probability distribution.

Here is a random integer between 1 and 100:

```
In[1]:= Random[Integer, {1, 100}]

Out[1]= 85
```

Here is a random integer in the range from −100 to 100:

```
In[2]:= Random[Integer, {-100, 100}]

Out[2]= -52
```

By omitting the range, the default of 0 to 1 is used. Here then, are ten random integers between 0 and 1:

```
In[3]:= Table[Random[Integer], {10}]

Out[3]= {0, 0, 0, 1, 1, 0, 0, 1, 0, 0}
```

Here is a real random number in the range from 1.1 to 5.2:

```
In[4]:= Random[Real, {1.1, 5.2}]

Out[4]= 2.35193
```

Here is a random real number that is in the default range of 0 to 1:

```
In[5]:= Random[Real]

Out[5]= 0.404758
```

In fact, by omitting the number type, we see that the default is for a real number with default range from 0 to 1:

```
In[6]:= Random[ ]

Out[6]= 0.0182791
```

Here is a complex random number. Notice that the real part is between 2 and 4 and that the imaginary part is between −3 and 5.

```
In[7]:= Random[Complex, {2 - 3 I, 4 + 5 I}]

Out[7]= 2.64459 + 4.19898 I
```

It should be noted that you can generate a random real number with n digits of precision by using Random[Real, {*range*}, *n*]. The following example produces a random real number between 0 and 3 using 40 digits of precision.

```
In[8]:= Random[Real, {0,3}, 40]

Out[8]= 1.717622610829332772114527522786499963327
```

Mathematica uses two different random number generators. One uses two cellular automaton rules to generate sequences, and the other method uses a "subtract-with-borrow" method with base 2^{31}. The interested reader should consult the references (Wolfram 1994, and Marsaglia and Zaman 1991) at the end of this appendix for a thorough treatment of these methods.

TESTING THE RANDOM NUMBER GENERATOR

There are numerous ways to look at and test the randomness of a random number generator. A rather intuitive method might look at the distribution of the numbers over the range from which they are taken.

First we create a list of 1000 integers, all taken between 1 and 10.

```
In[1]:= data = Table[Random[Integer, {1,10}], {1000}];
```

The two intuitive tests of randomness that we might be concerned with are the frequency with which each digit occurs, and the presence (or lack thereof) of any patterns in the sequence itself. If the above sequence contained approximately equal numbers of each digit but started off like the following sequence, we would be highly suspect.

$$0, 1, 2, 3, 4, 5, 6, 7, 8, 9, 0, 1, 2, 3, 4, 5, 6, 7, 8, 9, 0, \ldots$$

The frequency with which each digit occurs in the list data can be checked. First we load a package that contains the function Frequencies.

```
In[2]:= Needs["Statistics`DataManipulation`"]
```

```
In[3]:= ?Frequencies
```

Frequencies[list] gives a list of the distinct elements
   in list, together with the frequencies with which
   they occur.

```
In[4]:= Frequencies[data]
```

```
Out[4]= {{105, 1}, {90, 2}, {108, 3}, {95, 4}, {88, 5},
          {111, 6}, {99, 7}, {115, 8}, {94, 9}, {95, 10}}
```

Each of the digits occurs *roughly* 100 times, as expected. Of course, we wouldn't expect each digit to occur exactly 100 times, although on any given run of the data, this is certainly possible.

For a more careful analysis of random sequences, a variety of tests are available. One of the more commonly known tests for determining the effective randomness of a sequence is the χ^2 (chi-square) test. This test determines how evenly spread the numbers appear in a sequence and uses their frequency of occurrence to come up with a statistic. If n is the upper bound of a sequence of m positive numbers, then for a well-distributed random sequence, we would expect m/n copies of each number. The χ^2 test is given by the following formula:

$$\chi^2 = \frac{\sum_{1 \leq i \leq n}(f_i - m/n)^2}{m/n}$$

In the formula, f_i is the frequency of copies of i in the sequence. If the statistic is close to the upper bound n, then the numbers are assumed to be reasonably random. Here is an implementation of the χ^2 statistic:

```
In[5]:= ChiSquare[l_List] := Module[{m = Length[l], n = Max[l]},
           Sum[(Count[l, i] - m/n)^2, {i, 1, n}] / (m/n)]
```

Here is the χ^2 statistic for the above sequence of integers in the data list:

```
In[6]:= ChiSquare[data] //N
```

```
Out[6]= 7.66
```

In general, sequences of positive integers are considered sufficiently random if the χ^2 statistic is within $2\sqrt{n}$ of n. It is an interesting exercise to determine the χ^2 statistic for a variety of random number generators, such as the linear congruential method given above.

B.2 ∎ RANDOM NUMBER SEQUENCES

In many of the simulations in this book, it is often desirable to try to reproduce some behavior. For example, with many of these models, an initial configuration is generated by calling the Random function. If you wanted to repeat a particular simulation, but possibly increase the resolution (grid size), or bump up the number of values that each site can take on, then you will find yourself wanting to recreate random number sequences. This can be accomplished by *reseeding* the random number generator.

```
In[1]:= ?SeedRandom

SeedRandom[n] resets the pseudorandom number generator,
    using the integer n as a seed. SeedRandom[ ] resets
    the generator, using as a seed the time of day.
```

The random number generator is first seeded when you start up *Mathematica*. It uses the time of day when *Mathematica* is first started to generate the seed. By calling SeedRandom, you effectively reset the generator.

This seeds the random number generator and then gives five random real numbers between 0 and 1:

```
In[2]:= SeedRandom[4]
        Table[Random[], {5}]

Out[2]= {0.672212, 0.381237, 0.664015, 0.808308, 0.64736}
```

Reseeding the generator with the same seed causes the same sequence to be generated:

```
In[3]:= SeedRandom[4]
        Table[Random[], {5}]

Out[3]= {0.672212, 0.381237, 0.664015, 0.808308, 0.64736}
```

The current state of the random number generator can be checked by calling the function $RandomState. This function returns a big integer that represents the complete state of the random number generator at the moment that $RandomState is called.

In[4]:= **$RandomState**

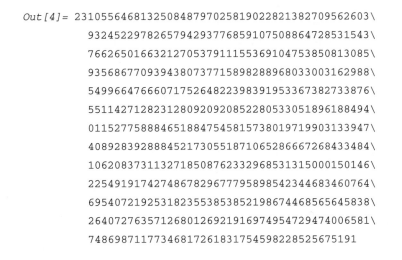

Out[4]= 23105564681325084879702581902282138270956260 3\
9324522978265794293776859107508864728531543\
7662650166321270537911155369104753850813085\
9356867709394380737715898288968033003162988\
5499664766607175264822398391953367382733876\
5511427128231280920920852280533051896188494\
0115277588846518847545815738019719903133947\
4089283928884521730551871065286667268433484\
1062083731132718508762332968531315000150146\
2254919174274867829677795898542344683460764\
6954072192531823553853852198674468565645838\
2640727635712680126921916974954729474006581\
7486987117734681726183175459822852567519 1

This is computed by going through an array of integers (representing the current state of the generator) and concatenates these integers to form the (base 2^{16}) digits of the big integer. It then computes some check sum digits to be able to verify that the big integer is in fact a valid $RandomState if a user uses it to set the random state at a later time.

■B.3 ■ USING DIFFERENT PROBABILITY DISTRIBUTIONS

BUILT-IN DISTRIBUTIONS

By default, *Mathematica* generates random number sequences from a uniform distribution. Random numbers can also be generated using any probability distribution. For example, if you wanted to generate samples from a normal (or Gaussian) distribution, this can be done by giving an argument to the Random function.

First we'll load a package that contains a variety of continuous probability distributions.

In[1]:= **Needs["Statistics`ContinuousDistributions`"]**

Here is a list of all of the distributions whose definitions are now loaded into memory:

In[2]:= **Names["Statistics`ContinuousDistributions`*"]**

```
Out[2]= {BetaDistribution, CauchyDistribution,
         ChiDistribution, ExponentialDistribution,
         ExtremeValueDistribution, GammaDistribution,
         HalfNormalDistribution, LaplaceDistribution,
         LogisticDistribution, LogNormalDistribution,
         NoncentralChiSquareDistribution,
         NoncentralFRatioDistribution,
         NoncentralStudentTDistribution,
         RayleighDistribution, UniformDistribution,
         WeibullDistribution}
```

THE NORMAL DISTRIBUTION

The normal distribution can be used to generate random number sequences. Here is the usage message for this distribution:

```
In[3]:= ?NormalDistribution
```

```
NormalDistribution[mu, sigma] represents the Normal
    (Gaussian) distribution with mean mu and standard
    deviation sigma.
```

Here is a function that uses the built-in Random to generate a random variable from the normal distribution with mean 0 and standard deviation equal to 1:

```
In[4]:= randNormal := Random[NormalDistribution[0,1]]
```

Here are eight samples using this distribution:

```
In[5]:= Table[randNormal, {8}]
```

```
Out[5]= {0.799800643079645, 1.810374350584248,
         0.2534409785918251, -2.259477769398556,
         -1.110348064994891, -1.306664601460623,
         -0.5276275338393789, 0.5148343517365129}
```

You can get a graphical display of the distribution by putting sample points into "bins" and then counting the number of points in each bin. This is done with the BinCount function.

First, we'll create a large set of samples.

```
In[6]:= data = Table[randNormal, {4000}];
```

This counts the number of sample points in the range from -4 to -3.9, then those from -3.9 to -3.8, and so on.

```
In[7]:= bindata = BinCounts[data, {-4, 4, .1}];
```

To display the distribution, we'll first need to load an auxiliary package.

```
In[8]:= Needs["Graphics`Graphics`"]
```

```
In[9]:= ?GeneralizedBarChart
```

```
GeneralizedBarChart[{{pos1, height1, width1}, {pos2,
    height2, width2},...}] generates a bar chart with the
    bars at the given positions, heights, and widths.
```

So, to put our data in the form that `GeneralizedBarChart` can use, we need triples of data, where each triple will consist of the number, its frequency, and the width of the bar that will represent that data point.

Using `Transpose`, we can pair up the numbers in the range of values with their respective frequencies.

```
In[10]:= Transpose[{Range[Floor[Min[data]] + .1,
                     Ceiling[Max[data]], .1],
                bindata}];
```

This shows a short list of this pairing:

```
In[11]:= Short[%]
```

```
Out[11]= {{-3.9, 0}, {-3.8, 0}, {-3.7, 0}, <<76>>, {4., 0}}
```

This prepares the data so that the pairs are turned into triplets, with each third element being set to 0.1. This will be the width of each bar in the bar chart ({pos_i, $height_i$, $width_i$}).

```
In[12]:= generalbindata =
            Transpose[{Range[Floor[Min[data]] + .1,
                         Ceiling[Max[data]], .1],
                    bindata}]; /.
              {a_, b_} -> {a, b, .1};
```

Finally, we can make the plot of the frequency:

In[13]:= **GeneralizedBarChart[generalbindata,**
          PlotRange -> All]

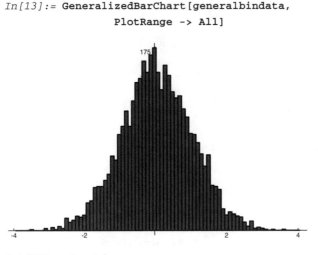

Out[13]= *-Graphics-*

These steps can be combined to produce one function that will plot a set of data by separating it into bins of length `incr`. Note the need to first load the necessary packages where the graphics and statistical functions are located.

In[14]:= **Needs["Graphics`Graphics`"];**
          Needs["Statistics`DataManipulation`"];

In[15]:= **PlotBinData[data_List, incr_]:=**
             Module[{min = Floor[Min[data]],
                    max = Ceiling[Max[data]]},
                 bindata = BinCounts[data, {min,max,incr}];
                 GeneralizedBarChart[
                     Transpose[{Range[min+incr, max, incr],
                               bindata}] /.
                     {a_, b_} -> {a, b, incr},
                     PlotRange -> All]
                 ]

Here is a typical call of the function using the original set of 4000 data points from above, this time separating the numbers into bins of length 0.2:

In[16]:= **PlotBinData[data, 0.2]**

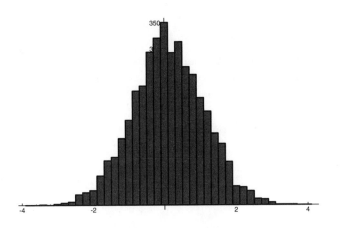

THE LOG NORMAL DISTRIBUTION

The following example shows how to use a log normal distribution with the random number generator.

In[17]:= **?LogNormalDistribution**

```
LogNormalDistribution[mu, sigma] represents the
    Log-Normal distribution with mean parameter mu and
    variance parameter sigma.
```

This function generates random numbers from the log normal distribution with mean 0 and variance 1.

In[18]:= **randLogNormal := Random[LogNormalDistribution[0,1]]**

This gives ten random numbers in the range from 0 to 1, using the log normal distribution.

In[19]:= **Table[randLogNormal, {10}]**

Out[19]= {0.2042030466727368, 1.394655549622289,
            0.6985550061427415, 3.104723846737352,
            2.927802630516859, 1.23815295334343,
            1.920008518669859, 0.759182114726928,
            0.6935792194239435, 0.7377646360861031}

Random numbers with a log normal distribution can be calculated much faster by using a normal distribution as follows:

```
In[20]:= lognormdist /:
            Random[lognormdist[mu_, sigma_]] :=
               N[Exp[mu + sigma Sqrt[-2 Log[Random[]]] *
                  Cos[2Pi Random[]]]]
```

Here are some timing comparisons:

```
In[21]:= Timing[Table[randLogNormal, {25}];]

Out[21]= {10.2333 Second, Null}
```

```
In[22]:= Timing[Table[Random[lognormdist[0, 1]], {25}];]

Out[22]= {0.116667 Second, Null}
```

ARBITRARY DISTRIBUTIONS

Any distribution can be used to generate sequences of random numbers. For example, you can use a probability density function of $\sin x$ to create a cumulative distribution and then use this distribution to generate random numbers.

The *cumulative distribution* at a variable x is the integral of the probability density function up to x. So in a general sense, differentiating the cumulative distribution function will give the probability density function.

The following example gives a cumulative distribution function in the range $[0, \pi]$ with the probability given by $\sin x$:

```
In[1]:= cdf[x_] = Integrate[Sin[t], {t, 0, x}]/
                     Integrate[Sin[t], {t, 0, Pi}]

Out[1]= 1 - Cos[x]
        ─────────
            2
```

Here, we generate the inverse function:

```
In[2]:= Solve[t == cdf[x], x]

Solve::ifun:
    Warning: Inverse functions are being used by Solve,
      so some solutions may not be found.

Out[2]= {{x -> ArcCos[1 - 2 t]}}
```

```
In[3]:= Inversecdf[t_] = x /. First[%]
```

```
Out[3]= ArcCos[1 - 2 t]
```

This inverse distribution function can now be used to generate random numbers.

```
In[4]:= Table[Inversecdf[Random[]], {20}]
```

```
Out[4]= {0.836316, 1.60252, 2.38008, 2.52242, 2.46135,
         0.174728, 0.723714, 1.58947, 1.20478, 2.22385,
         2.0048, 2.46512, 1.5573, 2.58463, 1.24751, 1.8066,
         2.02339, 2.48543, 1.16444, 0.170585}
```

REFERENCES

D. Knuth. *The Art of Computer Programming, Volume 2: Seminumerical Algorithms*, Second Edition. Addison-Wesley 1981.

G. Marsaglia and A. Zaman. A new class of random number generators. *The Annals of Applied Probability* 1 (1991) 462–480.

S. Wolfram. Random sequence generation by cellular automata. *Advances in Applied Mathematics* 7 (June 1986) 123–169.

Guide to Standard Mathematica Packages, Version 2.2. Technical Report. Wolfram Research, Inc., 1993.

*A*PPENDIX C

Computer Simulations and MathLink

by Todd Gayley

INTRODUCTION

As this book has demonstrated, *Mathematica* is an excellent environment in which to create simulation programs. The combination of a superb high-level programming language, an interpreted environment that encourages step-by-step development and testing, and rich visualization tools provide unparalleled ease of development. The only drawback is that *Mathematica* is demanding on system resources, both speed and memory. Fortunately, *Mathematica* has a facility called *MathLink* that makes it easy to integrate external functions written in a compiled language (C, for example). With *MathLink* it is possible to write external programs that become seamless extensions to the kernel. In this way, if you have an algorithm that needs to be written in C for efficiency reasons, you still have access to the rich *Mathematica* environment for all the auxiliary tasks—prototyping, preparing and managing input, analyzing and visualizing output, and so on.

This appendix describes how to use *MathLink* to call external functions from *Mathematica*. We start with a basic introduction to *MathLink*, then explore techniques in *MathLink* programming through the development of C source code for six of the simulations presented in the book. Most of the simulations enjoy a considerable speed increase when coded in C as external functions. This is particularly important for large simulations, where the running time can be reduced from hours to minutes. Even if you have no interest in *MathLink*, you will find these programs useful as fast versions of their respective simulations.

We assume a basic familiarity with C, and since our focus is on using *MathLink*, we will not deal extensively with the actual implementations of the algorithms, except to highlight *MathLink* programming issues and contrast programming styles in *Mathematica* and C.

All the source code is included on disk, along with ready-to-run executables for Macintosh and Windows. In addition, there is a sample notebook that shows how to install and use the programs. The source code is ANSI C, and it will compile without modification on any platform *Mathematica* supports. Instructions for building and running the programs appear below.

The code presented here compiles under version 2.2.2 of *MathLink* (the current shipping version on all platforms as of this writing) and later versions. There is only one line that is not compatible with earlier versions (back to version 2.1); it is noted where it appears, and it can be deleted without materially affecting the program. This is an issue only if you want to compile the programs yourself; the supplied Macintosh and Windows executables will work with any version of *Mathematica* that supports *MathLink*.

C.1 ■ INTRODUCTION TO *MATHLINK*

It is not our intention here to provide an in-depth tutorial on *MathLink*. The main documentation for *MathLink* is the *MathLink Reference Guide*, which ships with most versions of *Mathematica* and can also be obtained on *MathSource* or ordered from Wolfram Research directly. In addition, there is an extensive *MathLink* tutorial on *MathSource*, as well as various sample programs. You should consult these materials for more information about *MathLink* programming.

MathLink is a library of functions that implement a protocol for sending and receiving *Mathematica* expressions. The easiest and most common application is to allow external functions written in other languages to be called from within the *Mathematica* environment. This is how we use *MathLink* in writing these simulations. As we shall see, it is a relatively simple matter to incorporate such routines into *Mathematica*.

We refer to external functions that are called from *Mathematica* as *installable* functions, since they use the `Install` function to be incorporated into the *Mathematica* environment. For each external function you want to call from *Mathematica*, you write a *template* that specifies information about the function, including its name, the arguments that it needs to be passed and their types, and the type of argument it returns. This template file is then passed through a tool called **mprep**, which writes C code that manages most of the *MathLink*-related aspects of the program. The magic here is in this processing by **mprep**. It writes a lot of C code that "wraps around" your function and handles a lot of complexity you'd rather not have to confront. It also lets you write portable programs, because all the machine-specific code is written by **mprep** according to the particular platform it is running on.

The first program we develop, the two-dimensional random walk, is a bit too complicated to serve as the introduction to *MathLink*. Thus, we begin with the venerable `addtwo` program, one of the sample programs included with *MathLink*. As its name implies, this function simply adds two integers. Here is how you would write such a function in C:

```
int addtwo(int i, int j) {
    return(i + j);
}
```

Now that we have a function that takes two integers and returns their sum, how do we go about making it installable? The first thing we need is a `main` function. Every C program must have a `main` function, which serves as the program's entry point. The `main` function for an installable program is just "boilerplate" code that you copy and paste out of a sample program. Here is the (one-line) `main` function:

```
int main(int argc, char *argv[]) {
    return(MLMain(argc, argv));
}
```

As you can see, it just calls the "real" `main` function `MLMain`, which is written by **mprep**. For Windows programmers, `main` is slightly more complicated, but you don't have to worry about what it means—just paste it into your program. That's all the C code we need to write for our `addtwo` program. What's left is to create the template file for the function. This is the specification that **mprep** will use to write the *MathLink*-related portions of the program. Template files are given the `.tm` extension, so we would call this file `addtwo.tm`:

```
:Begin:
:Function:       addtwo
:Pattern:        AddTwo[i_Integer, j_Integer]
:Arguments:      { i, j }
:ArgumentTypes:  { Integer, Integer }
:ReturnType:     Integer
:End:
```

The `:Function:` line specifies the name of the C routine that will be called. The `:Pattern:` line shows how the routine will be called from *Mathematica*. The pattern given on this line will become the left-hand side of a function definition, exactly as you would type it if you were creating the entire function in *Mathematica*. The `:Arguments:` line specifies the expressions to be passed to the external program. This line need not merely echo the variable names on the `:Pattern:` line, although it often will. You could put any *Mathematica* code there—for example, `{i^2, j-1}`. What you put on the `:Pattern:` line and the `:Arguments:` line will be used to create a function definition that could be caricatured as follows:

```
AddTwo[i_Integer, j_Integer] :=
    SendToExternalProgramAndWaitForAnswer[{i, j}]
```

The `:ArgumentTypes:` and `:ReturnType:` lines contain keywords used by **mprep** to create the appropriate *MathLink* calls that transfer data across the link. The **mprep** program will read this `.tm` file and produce a C file with the `.tm.c` extension.

You might put the few lines of C code into a file named `addtwo.c`, but a handy feature of **mprep** is that it will pass through any C code in a template file unmodified, so you can put the C and template(s) into a single `.tm` file. This is what we have done with all the simulation programs we develop. Here's what the final `addtwo.tm` looks like:

```
:Begin:
:Function:      addtwo
:Pattern:       AddTwo[i_Integer, j_Integer]
:Arguments:     { i, j }
:ArgumentTypes: { Integer, Integer }
:ReturnType:    Integer
:End:

int addtwo(int i, int j) {
    return(i + j);
}

int main(int argc, char *argv[]) {
    return(MLMain(argc, argv));
}
```

The details of building the executable from the `addtwo.tm` source file differ from platform to platform. On Unix, you can use the **mcc** script that comes with *Mathematica*:

```
mcc addtwo.tm -o addtwo
```

mcc is a simple utility that does nothing more than run **mprep** on any `.tm` files, to create `.tm.c` files, and then send all the source files through the **cc** compiler. Normally, the `.tm.c` file is deleted by **mcc** after it has been compiled, but if you want to look at it you can prevent its deletion by specifying the `-g` command-line option to **mcc**. Perusing this file is not for the C novice, but it is very useful for understanding what goes on in installable programs. On Macintosh and Windows, the steps to build the program will be different, and you should consult the README file that comes with *MathLink*. Typically, you will open a "project" in your development environment, add the `.tm.c` file (you may have to run **mprep** on the `.tm` file manually), the `mathlink.h` header file, and the *MathLink* library, along with some other platform-specific libraries, then build the program.

To use the `AddTwo` function in *Mathematica*, you launch the external program with the `Install` function:

```
In[1]:= link = Install["addtwo"]

Out[1]= LinkObject[addtwo, 2, 2]
```

The function `LinkPatterns` shows what functions are defined by the external program associated with a given link:

```
In[2]:= LinkPatterns[link]

Out[2]= {AddTwo[i_Integer, j_Integer]}


In[3]:= AddTwo[2,3]

Out[3]= 5
```

You'll notice that in writing the program, we didn't have to make any *MathLink* calls; everything was handled by the code written by **mprep**. That code gets the two integer arguments from *Mathematica* and sends back the sum returned by our function. It is convenient to have this done automatically for us, but in many cases we will want to do some of these steps ourselves. In particular, it is often useful to manually send the result back to *Mathematica*. We tell **mprep** that we will send the result back by using the keyword `Manual` on the `:ReturnType:` line of the template. Here is what the template and `addtwo` function would look like:

```
:Begin:
:Function:       addtwo
:Pattern:        AddTwo[i_Integer, j_Integer]
:Arguments:      { i, j }
:ArgumentTypes:  { Integer, Integer }
:ReturnType:     Manual
:End:

void addtwo(int i, int j) {
    MLPutInteger(stdlink, i + j);
}
```

Notice the change in the prototype for the `addtwo` function. It no longer returns anything; rather, we manually make the `MLPutInteger` call that sends the sum back to *Mathematica*.

There are two common reasons for doing this. One is that we may want to send back an expression that is not one of the set handled automatically by **mprep** (limited to `Integer`, `Real`, `String`, and `Symbol`). The other case is that we may need to send different types of things in different circumstances. For example, if there is some problem during execution (*e.g.*, the failure of a memory request) we want to send the symbol `$Failed`, whereas if the function completes normally we may be sending back an array of integers. In every one of our simulation programs, we use `Manual` as the return type for both of these reasons.

Another important technique is returning a *Mathematica* function from our program. A very common type of function that you might need to return is `List`, although you might not think of this as a function, because it is usually written in a shorthand form with curly braces, as in {1, 2, 3}. To send functions across *MathLink*, we use `MLPutFunction`. As an example, let's say we want the `addtwo` function to return not just the sum, but rather a list of the two original integers and the sum. Here's how that could be done:

```
void addtwo(int i, int j) {
    MLPutFunction(stdlink, "List", 3);
    MLPutInteger(stdlink, i);
    MLPutInteger(stdlink, j);
    MLPutInteger(stdlink, i + j);
}
```

Note that `MLPutFunction` takes as arguments the head of the function (as a string), followed by the number of arguments (*i.e.*, the length of the list). It is important to note that, at the time you send the head of the function, you need to know how many arguments will follow. This requirement will influence our strategies for returning the results from our simulations.

C.2 ■ COMPILING AND RUNNING THE SIMULATIONS

MACINTOSH

There are ready-to-run executables for all the programs provided on disk. We also include THINK C (version 6) project files for each simulation, to simplify the process of building the programs if you want to modify them in any way. If you have THINK C, and you have obtained and installed the *MathLink* Developer's Kit materials, you can just double-click on any project file and be instantly ready to edit and recompile the corresponding program. If you don't have THINK C, you can simply use the source files in your favorite development environment. For each program, there is just one source file, a `.tm` file that must be processed with **mprep** (or **SAmprep**, for non-MPW users) into a `.tm.c` file, which is what will be sent through the C compiler. You should consult the README file that comes with the *MathLink* Developer's Kit for detailed information about building installable programs with your development environment.

To use a program in *Mathematica*, you use the `Install` function:

```
In[1]:= link = Install["'Hard Disk:My Folder:phasesC'"]
```

Note that the colon is used to separate folder names in Macintosh pathnames, and that we need an inner set of single quotes around the pathname if it contains any spaces. You can either put the programs in a folder where they will be found by the `Install` function (the *Mathematica* folder, for example) or simply specify the full pathname to the file, as in the above example.

Each of the external functions has the same name as the corresponding *Mathematica* function, but with a "C" appended. Thus, the two-dimensional walk function is called from *Mathematica* as `Walk2DC`. Use `LinkPatterns` to find out what functions are available in each program:

```
In[2]:= LinkPatterns[link]

Out[2]= {PhasesC[s_Integer, n_Integer, maxIterations_Integer],
            SeedRandomC[seed_Integer]}
```

WINDOWS

There are ready-to-run Windows executables for all six programs on disk. If you want to modify and compile the programs yourself, you will need the *MathLink* Developer's Kit for Windows and any commercial C compiler capable of creating Windows programs. The executables we supply were built with Borland C++ 4.0, and we have included the "project" (`.ide`) files that were used for each one. If you have Borland C++ you can use these same `.ide` files, but you will have to change some pathnames to reflect the locations of the relevant files on your system. The programs on disk are built as 16-bit programs, which in some cases imposes limitations on the size of simulations that can be run. For example, if the simulation involves a square cell matrix, then it cannot be bigger than 256 by 256. If you need to go beyond these limits, then you can recompile the programs as Win32 programs.

Each of the simulation programs is written as a single `.tm` file. You will need to run **mprep** on them to create `tm.c` files. Here's an example DOS command line line to invoke **mprep** (which is part of the *MathLink* Developer's Kit):

```
c:\mathlink\bin\mprep  phases.tm -o phasestm.c
```

What you name the `tm.c` file is up to you. Note that in Windows, where filenames cannot have multiple periods in them, we usually choose filenames of the form `*tm.c`. Once you have created a `tm.c` file, include it as the single C file in your project. There is more detailed information about building installable programs under Windows in the README file that comes with the *MathLink* Developer's Kit for Windows, including specific examples using Microsoft Visual C++, Borland C++, and Symantec C++.

To use a program in *Mathematica*, you use the `Install` function:

```
In[1]:= link = Install["c:\\mydir\\phasesC"]
```

Note that you need to use double backslashes to indicate directory separators inside *Mathematica* strings. You can either put the programs in a directory where they will be found by the `Install` function (the *Mathematica* directory, for example) or simply specify the full pathname to the file, as in the above example.

Each of the external functions has the same name as the corresponding *Mathematica* function, but with a "C" appended. Thus, the two-dimensional walk function is called from *Mathematica* as `Walk2DC`. Use `LinkPatterns` to find out what functions are available in each program:

```
In[2]:= LinkPatterns[link]

Out[2]= {PhasesC[s_Integer, n_Integer, maxIterations_Integer],
          SeedRandomC[seed_Integer]}
```

UNIX

Unix users will probably find that the entire build process can be accomplished by invoking the **mcc** script, as in:

```
mcc phasesC.tm -o phasesC
```

If you get syntax errors from the compiler, then it is likely that the **cc** compiler on your machine is not ANSI-compliant. You will need to either modify the **mcc** script so that it calls an ANSI-compliant compiler (*e.g.*, **gcc**, the GNU C compiler), or make the program manually. The first step is to run **mprep**:

```
/math/Bin/MathLink/mprep phasesC.tm -o phasesC.tm.c
```

Now, pass the `.tm.c` file to your compiler, specifying the location of the `mathlink.h` header file (the `math/Source/Includes` directory) and the *Math-Link* library (the `math/Bin/MathLink` directory).

To use the program in *Mathematica*, you use the `Install` function:

```
In[1]:= link = Install["/mydir/phasesC"]
```

You can either put the programs in a directory where they will be found by the `Install` function (the *Mathematica* directory, for example) or simply specify the full pathname to the file, as in the above example.

Each of the external functions has the same name as the corresponding *Mathematica* function, but with a "C" appended. Thus, the two-dimensional walk function is called from *Mathematica* as `Walk2DC`. Use `LinkPatterns` to find out which functions are available in each program:

```
In[2]:= LinkPatterns[link]

Out[2]= {PhasesC[s_Integer, n_Integer, maxIterations_Integer],
            SeedRandomC[seed_Integer]}
```

C.3 ■ DEVELOPING THE SIMULATIONS

The simulation programs presented in this book are good candidates to be rewritten as *MathLink* external functions because they are very computation-intensive. A modest-sized simulation can take several minutes to complete, and larger runs can take hours. Unfortunately, these simulations also have a feature that makes them less than ideal for *MathLink*: they return large quantities of data to *Mathematica*. Although the speed of transmission over *MathLink* is relatively fast, it still creates a bottleneck that prevents the full realization of the speed that C provides for the actual execution of the simulations. For some of these programs, the time taken by running a generation of the algorithm is essentially instantaneous compared to the time it takes to send the results over *MathLink*, assemble them into a *Mathematica* expression, and send that expression through the evaluator. Many of the *MathLink* tricks and techniques we discuss will be devoted to optimizing these steps. *MathLink* is constantly being improved, and some of the efficiency concerns that we code around will disappear in future releases.

The fact that most of the time is taken in returning the results to *Mathematica* implies that careful optimization of the algorithms is of limited value. Accordingly, the algorithms have been coded in a straightforward manner that emphasizes simplicity. Nevertheless, for stylistic reasons we have included some standard C optimization techniques even when they might detract from readability. We do not claim that these programs are "optimal" in any way, and we invite the reader to experiment with alternative methods.

It is a worthwhile exercise to rewrite one of the simulations from this book in C. There are few tasks that will give you a clearer appreciation for the *Mathematica* language and environment. A few of the features that you give up when writing in C are cited below.

Interpreted environment. Possibly the single feature of *Mathematica* that most increases productivity is its interpreted environment, which allows you to build and test programs one step at a time. It encourages creativity and experimentation, and greatly simplifies debugging. The edit-compile-crash cycle of compiled languages makes every minor revision a time-consuming operation, and the need to write a complete program to test every new idea makes it difficult to follow up on alternative insights.

High-level programming constructs. Consider all the steps you need to perform to duplicate the functionality of `Map`, `FoldList`, or `FixedPointList`. Each of these is a significant program in itself when written in C. Having access to high-level functions like these saves countless lines of code.

Diverse programming styles. *Mathematica* allows many different styles of programming: procedural (like C), functional, rule-based, object-oriented, and so on. Many programs that are difficult to write in one style become much simpler in another, so it is convenient to have easy access to a variety of styles.

Automatic memory management. *Mathematica* completely isolates the programmer from issues of memory allocation and deallocation. Improper memory management is a source of a great number of C programming bugs, and these bugs are notoriously difficult to detect and track down.

Typeless variables and structures. Variables in *Mathematica* can hold objects of any type, and lists can hold any type or mixture of types. If you need a data structure that holds, say, pairs of strings and numbers, it is trivial to implement as a nested list. This flexibility makes it easy to construct and handle data structures that in C would have to be rigidly defined and painstakingly manipulated.

A WORD ABOUT RANDOM NUMBERS

The generation of random numbers is an important part of these simulations. Random numbers are a classic subject in computer science, and a lot has been written about the issue, including elsewhere in this book. For simplicity, we rely on the `rand()` function, a standard part of ANSI C. The quality and properties of this generator vary from platform to platform (for example, on the Macintosh it is usually implemented by calling the Toolbox function `Random`). In general, `rand()` is not a particularly good generator; we maintain, however, that it is sufficient for our needs. If you would like to use a different function, feel free to change the calls to `rand()`.

One property of `rand()` as it is implemented on most machines is that the last few bits in the number are highly nonrandom. For example, the sequence alternates even and odd numbers (which means that the last bit alternates between 0 and 1). This is a problem if the output of `rand()` is being used naively as a yes/no toggle. Similarly, there are cases where we need to choose between some small number of states, say four, so we might use `rand() % 4` (`%` means "mod" in C). This tests the last two bits in the number. Instead of using the last few bits, we want to use bits in the middle of the number, and to do that we simply "shift right" the number, so that some bit in the middle, say the tenth, now becomes the rightmost bit. Therefore, you will see constructions like `(rand()>>10) % 4` instead of `rand() % 4`.

C.4 ■ C PROGRAMS

WALK2DC

The two-dimensional random walk is a simple algorithm, making it a good place to start in our explorations. Our function, `Walk2DC`, will take an integer n as its argument, which specifies the length of the walk. The result we put back to *Mathematica* will be a list of length $n + 1$ (including the starting point) of x-y pairs. To decide which direction to move in each step, we need to select among four states: North, East, West, and South. This is accomplished with a `switch` statement that tests the value of a random number mod 4. If x and y are variables that hold the x and y positions of the walk, we can use the following to generate the n steps:

```
for(i = 1; i <= n; i++) {
    switch((rand()>>10) % 4) {
        case 0:  x++; break;
        case 1:  y++; break;
        case 2:  x--; break;
        case 3:  y--;
    }
}
```

This function has the nice feature that we know in advance how long the resulting list will be ($n + 1$ elements). As discussed earlier, whenever you call `MLPutFunction`, you need to specify the number of arguments the function will have. In our simulations, the function we will be sending back is `List`, and knowing ahead of time how long the resulting list will be allows us to send each element as it is generated, rather than accumulating the entire result in memory and sending it at the end.

At this point, a few comments in the code should sufficiently explain the complete function:

```
void walk2D(int n) {

    int i,
        x,
        y;

    x = y = 0;   /* Starting values of x and y positions */

    /* We put the head List right away */
    MLPutFunction(stdlink, "List", n + 1);
```

```
/* Each element of the result list is itself a list of x
   and y values. Here we put the first element, {0, 0} */
MLPutFunction(stdlink, "List", 2);
MLPutInteger(stdlink, x);
MLPutInteger(stdlink, y);

for(i = 1; i <= n; i++) {
    switch((rand()>>10) % 4) {
        case 0:  x++; break;
        case 1:  y++; break;
        case 2:  x--; break;
        case 3:  y--;
    }
    MLPutFunction(stdlink, "List", 2);
    MLPutInteger(stdlink, x);
    MLPutInteger(stdlink, y);
}
}
```

Here is the template:

```
:Begin:
:Function:       walk2D
:Pattern:        Walk2DC[n_Integer]
:Arguments:      {n}
:ArgumentTypes:  {Integer}
:ReturnType:     Manual
:End:
```

If we `Install` and run this program, we see that it is no faster than the *Mathematica*-only implementation! This shows that the bottleneck in sending a long list back to *Mathematica* can overshadow any speed gained from running the algorithm in C. The moral is that if your external function needs to send a lot of data back to *Mathematica*, you might want to test how long it takes to send that much data before you actually go about implementing the algorithm. Here, the algorithm itself is instantaneous compared with the time it takes to send the data back to *Mathematica*. In later simulations we will see the effects of this bottleneck, but the advantage of using *MathLink* will nevertheless become considerable, since the algorithms run slowly in *Mathematica*.

Rather than sending back the data as it is generated, we could also accumulate it in memory and send it back only after the function runs to completion. This is especially convenient because *MathLink* has functions for transferring arrays of numbers in a single step. The drawback is that we have to deal with memory allocation and deallocation. Let's see how the function would look using this method.

The basic technique is to allocate an array big enough to hold the entire result, then fill it up as we loop through the simulation. An important aspect of C that is used throughout the simulations is the way pointer arithmetic is performed. When you add a number n to a pointer, the pointer is incremented not by n, but rather by n times the size of the object that the pointer points to. For example, if `ptr` is declared as a pointer to `short int`, then `ptr++` (which adds 1 to `ptr`) changes the address not by one byte but by two bytes (1 times the size of a short). This makes it very easy to march through an array performing some operation on each element; ptr++ always makes ptr point to the next element in the array, no matter how big the elements are.

The *MathLink* function we will use to send a multidimensional array back to *Mathematica* in a single step is MLPutShortIntegerArray. In addition to the link pointer, the arguments to MLPutShortIntegerArray are: the short array (that is, the address of its first element); a dims array, which is an array of longs that specify the length in each dimension; a heads parameter, which specifies the heads of the expression in each dimension (if you simply pass in NULL here, it assumes the default of List as the head at each level); and the number of dimensions.

```c
void walk2D_array(int n) {

    int    i;
    short  x,
           y,
           *index,
           *array;
    long   dims[2];
    dims[0] = n + 1;
    dims[1] = 2;

    /* Here we allocate memory to hold the entire array */
    array = malloc((n + 1) * 2 * sizeof(short));
    /* The memory allocation might fail, and if so we want to
       bail out of the function immediately. */
    if(!array) {
        MLPutSymbol(stdlink, "$Failed");
        return;
    }

    x = y = array[0] = array[1] = 0;
    index = array + 2;  /* for the first x=0, y=0 */
    for(i = 1; i <= n; i++) {
        /* index is the pointer into the array. A statement like
           *index++ = x means "put the value in x at the address
           pointed to by index, then increment index" (so it
           points to the next element in the array) */
        switch((rand()>>10) % 4) {
            case 0:  *index++ = ++x;
                     *index++ = y;
                     break;
            case 1:  *index++ = x;
                     *index++ = ++y;
                     break;
            case 2:  *index++ = --x;
                     *index++ = y;
                     break;
            case 3:  *index++ = x;
                     *index++ = --y;
        }
    }
    MLPutShortIntegerArray(stdlink, array, dims, NULL, 2);
    free(array);
}
```

This function has the feature that it returns different types of results in different situations. If the request for memory to hold the array fails, it returns the symbol $Failed; otherwise, it runs the simulation and returns a list of *x-y* pairs.

This version of the function is more complicated than the first, and it runs at essentially the same speed. However, a future version of *MathLink* (after Version 2.2.3) will have optimizations that allow the MLPutShortIntegerArray to execute much more quickly than putting each sublist and integer "by hand," as was done in the first version.

SEEDRANDOMC

Because every one of the simulations uses the random number generator, it is useful to have a way to seed it. Thus, every program also defines a second external function named SeedRandomC, which will seed the random number generator. The function takes an integer argument that will be passed directly to the ANSI C function srand(). Here is the template and C code for SeedRandomC:

```
:Begin:
:Function:       seed_random
:Pattern:        SeedRandomC[seed_Integer]
:Arguments:      {seed}
:ArgumentTypes:  {Integer}
:ReturnType:     Manual
:End:

void seed_random(int seed) {
    srand((unsigned int) seed);
    MLPutSymbol(stdlink, "Null");
}
```

We emphasize that this function is not in a separate program that you need to Install; every one of the simulation programs has its own version of SeedRandomC.

EPIDEMICC

The epidemic function has a somewhat complicated algorithm when coded in C, but we present it second because it is relatively simple in its use of *MathLink*. While some of the simulations in this book take place on a fixed-size grid, where each site needs to be checked every generation to see if its state must be updated, others are concerned with only one site and its neighbors in any generation. In these cases, rather than keeping track of a "board" of sites, we instead keep track of various lists of sites; *e.g.*, the Epidemic model has three lists, cluster, perimeter, and reject whose sizes change during the course of the simulation.

There are many ways to implement lists of arbitrary size that grow and perhaps also shrink during execution. We need to efficiently add elements to a list, delete elements (only for the perimeter list), and most importantly, test whether a given element is a member of a list. Standard techniques for implementing such collections in C include linear linked lists, binary trees, hash tables, and other more exotic schemes. Rather than bog down our code in the implementation details of these schemes, we choose a more straightforward method. We simply allocate an array of sites, initially empty, then fill it with sites of all three states as the simulation proceeds. As new sites are "touched" they are added to the end, and as old sites change state they are updated in place. With a little careful bookkeeping, at most two passes through the array are needed in each generation. If this program were used for large simulations (say, greater than a few thousand sites), you would want to choose a more sophisticated data structure (a binary tree, perhaps) and implement the three lists separately.

We need to make a memory allocation at the beginning of the program (with `malloc`), and of course any memory request can fail, so we need to check whether it succeeds. What should we do if it fails? The easiest thing is just to send the symbol `$Failed` to *Mathematica* and return, and this is what we do here and in the other simulations. A more sophisticated action would be to trigger a message in *Mathematica* alerting the user to what had happened, but this is beyond the scope of our treatment of *MathLink*.

Another new feature introduced in this simulation is the ability to abort an external function. Any time you write a function that might take an appreciable amount of time to run, you should allow the user to abort it. If the user signals an abort (command-period on Macintosh, control-C on Unix, etc.) while an external program is executing, *Mathematica* sends the program an abort "message." The low-level handling of these messages is taken care of for you by the code that **mprep** writes. You only need to know that installable programs have a global variable named `MLAbort` whose value will reflect whether the user has requested an abort. It is 0 by default, but will get set to 1 if an abort is requested. You should check this variable periodically to determine whether to continue the calculation or not. Our simulations all have a main "generation" loop, and it is a simple modification to check `MLAbort` in each trip through the loop.

What should your function do if it detects that `MLAbort` has been set to 1? One possibility is to immediately send back the symbol `$Aborted` and return. A better method is to send back the function `Abort[]`. This will cause the entire computation to abort in the case where the call to your external function is part of a larger computation in *Mathematica*. This mimics the behavior of *Mathematica* functions: If a user evaluates `f[g[x]]` and aborts during the execution of `g[x]`, the calculation returns simply `$Aborted`, not `f[$Aborted]`.

There is one more issue in making a function abortable. On systems without preemptive multitasking (*i.e.*, Macintosh and Windows), your function must period-ically yield the processor so that *Mathematica* actually gets a chance to detect and send the abort to you. *MathLink* handles this issue by incorporating what is known as a "yield function," which is a function that is called internally by *MathLink* at certain times to yield the processor to other programs. (On the Macintosh, for example, it calls the Toolbox function `WaitNextEvent`.) *MathLink* supplies a default yield function, and it also lets you specify one of your own, although that is beyond the scope of this discussion.

One solution to the yielding problem is to call *MathLink*'s yield function periodically. In the `EpidemicC` program, we do this once every generation. You manually call the yield function using `MLCallYieldFunction`. This allows you to write portable code and not have to delve into the specific Macintosh and Windows system calls needed to appropriately yield the processor. Take a look at the code for the exact syntax, and don't worry about what the arguments mean—just copy the line verbatim into your own programs. `MLCallYieldFunction` is a new addition to *MathLink*; if your linker gives you "undefined function"-type errors, you do not have a recent enough version of *MathLink*. In this case, you can just comment out the line.

Calling the yield function is unnecessary in two situations. The first is when your program is running under Unix, since the operating system takes care of switching the CPU among processes. The second situation is when you are periodically making *MathLink* calls to send results back (as in the `Walk2DC` program). In that case, *MathLink* will be calling the yield function itself, so you don't have to worry about it. In other words, you only have to call the yield function if you are not making any calls to *MathLink* functions that themselves will call the yield function. In the `EpidemiC` program, we save all the results until the end and send them back as a unit, so we need to yield the processor manually. Later simulations will be sending back results in each generation, so there is no need for direct calls to the yield function.

Here is the template for the `EpidemicC` function:

```
:Begin:
:Function:      epidemic
:Pattern:       EpidemicC[max_Integer, p_Real]
:Arguments:     {max, p}
:ArgumentTypes: {Integer, Real}
:ReturnType:    Manual
:End:
```

Here is the C code:

```
#include "mathlink.h"
#include <stdlib.h>

#define TRUE 1
#define FALSE 0
```

```
#define MAX_SITES   20000  /* Note the hard-coded limit. This is to
                               avoid memory allocations of >64 Kb in
                               Windows. If using Macintosh or Unix
                               (or a Win32 program under Windows),
                               you can increase this number */

#define CLUSTER     1
#define PERIMETER   2
#define REJECT      3

struct site {
             char xcoord;
             char ycoord;
             char type;
};

void epidemic(int max_size, double p);

/**************  epidemic  ***************/

void epidemic(int max_size, double p) {

    int    i,
           perimeter_length,
           cluster_length,
           sites_length,
           chosen_site,
           north_empty,
           south_empty,
           east_empty,
           west_empty;

    char   x,
           y,
           y_north,
           y_south,
           x_east,
           x_west;

    struct site  *sites,
                 *site_ptr;

    sites = (struct site *) malloc(MAX_SITES * sizeof(struct site));
    if(!sites) {
        MLPutSymbol(stdlink, "$Failed");
        return;
    }

    /* Set up the first 5 sites */
    sites[0].xcoord = 0;
    sites[0].ycoord = 0;
    sites[0].type = CLUSTER;
    sites[1].xcoord = 1;
    sites[1].ycoord = 0;
    sites[1].type = PERIMETER;
    sites[2].xcoord = 0;
    sites[2].ycoord = 1;
    sites[2].type = PERIMETER;
    sites[3].xcoord = -1;
    sites[3].ycoord = 0;
    sites[3].type = PERIMETER;
    sites[4].xcoord = 0;
    sites[4].ycoord = -1;
    sites[4].type = PERIMETER;
```

```
cluster_length = 1;
perimeter_length = 4;
sites_length = 5;

while(perimeter_length > 0 &&
        cluster_length < max_size &&
            sites_length <= MAX_SITES - 3 &&
                !MLAbort) {

    /* Choose a random perimeter site */
    chosen_site = rand() % perimeter_length + 1;

    /* For efficiency, we start at end of sites array and count
        backward, since perimeter sites are mainly clustered at
        the end of the array */

    site_ptr = sites + sites_length;
    for(i = 1; i <= chosen_site; i++) {
        do site_ptr--; while(site_ptr->type != PERIMETER);
    }
    if((1.0 * rand())/RAND_MAX > p) {
        site_ptr->type = REJECT;
        perimeter_length--;
    } else {
        site_ptr->type = CLUSTER;
        cluster_length++;
        perimeter_length--;
        x = site_ptr->xcoord;
        y = site_ptr->ycoord;
        y_north = y + 1;
        y_south = y - 1;
        x_east = x + 1;
        x_west = x - 1;
        north_empty = south_empty = east_empty =
        west_empty = TRUE;

        /* We make one pass through the array looking for
            non-empty neighbor sites; otherwise, neighbor
            sites are added as new perimeter sites. */
        for(site_ptr = sites, i = 0; i < sites_length; i++,
            site_ptr++) {
            char temp_x = site_ptr->xcoord,
                temp_y = site_ptr->ycoord;
            if(temp_x == x)
                if(temp_y == y_north)       north_empty = FALSE;
                else if(temp_y == y_south)  south_empty = FALSE;
            if(temp_y == y)
                if(temp_x == x_east)        east_empty = FALSE;
                else if(temp_x == x_west)   west_empty = FALSE;
        }
```

```
            if(north_empty) {
                site_ptr->xcoord = x;
                site_ptr->ycoord = y_north;
                site_ptr->type = PERIMETER;
                site_ptr++;
                perimeter_length++;
                sites_length++;
            }
            if(south_empty) {
                site_ptr->xcoord = x;
                site_ptr->ycoord = y_south;
                site_ptr->type = PERIMETER;
                site_ptr++;
                perimeter_length++;
                sites_length++;
            }
            if(east_empty) {
                site_ptr->xcoord = x_east;
                site_ptr->ycoord = y;
                site_ptr->type = PERIMETER;
                site_ptr++;
                perimeter_length++;
                sites_length++;
            }
            if(west_empty) {
                site_ptr->xcoord = x_west;
                site_ptr->ycoord = y;
                site_ptr->type = PERIMETER;
                site_ptr++;
                perimeter_length++;
                sites_length++;
            }
        }

        /* Call the MathLink yield function manually.
           This makes our program friendly in the cooperative
           multitasking environment of Macintosh and Windows,
           and also allows abort requests to be processed */

        MLCallYieldFunction(MLYieldFunction(stdlink), stdlink,
        (MLYieldParameters)0);
    }

    if(MLAbort || sites_length > MAX_SITES - 3) {
        MLPutFunction(stdlink, "Abort", 0);
    } else {
        MLPutFunction(stdlink, "List", cluster_length);

        /* March through sites array, sending back coords of
           CLUSTER sites */
        for(site_ptr = sites, i = 0; i < cluster_length; i++,
            site_ptr++) {
            while(site_ptr->type != CLUSTER) site_ptr++;
            MLPutFunction(stdlink, "List", 2);
            MLPutInteger(stdlink, site_ptr->xcoord);
            MLPutInteger(stdlink, site_ptr->ycoord);
        }
    }
    free(sites);
}
```

ONELANEC

The simulations so far have shared the feature that they don't send a huge amount of data back to *Mathematica*. They build up a modest-sized result list and send it back all at once, rather than sending back a big array of data every generation. The OneLaneC simulation, however, sends back an array every generation, so it is a candidate for special efforts to optimize the *MathLink* transmission. However, due to the nature of the data that it sends, there is little that can be done in the way of optimization tricks. Specifically, the problem is that we send back mixed symbolic and integer data (the symbol e is used for an empty space in the "road"). This requires us to use a series of low-level calls (MLPutInteger, MLPutSymbol) to send the data one element at a time. If we wanted to use a different format for the results (*e.g.*, −1 instead of e for an empty space), then some optimizations become available, along the same lines as those used in later simulations.

The logic of the algorithm implemented here is straightforward. We start at some location in the road array (the beginning of the array is a logical place). We step forward until we find the first car, marking all the spaces we step over as empty. These first spaces are special in that they may get filled later on, by a car that wraps around from the end of the array. When we reach a car, we look ahead to see how many spaces separate the next car. From this we determine the new speed, and move the car forward that many spaces, marking all the spaces we step over as empty. These steps are repeated for all cars.

This is the first simulation that has a "board" of some sort that is updated each generation. In all such simulations we need to allocate only enough memory to hold two copies of the board: one "read-only" board for the current generation, and one "write-only" board that is filled in for the next generation. After the next-generation board is complete, we send it to *Mathematica*, then simply exchange the pointers to the two boards so that the next-generation board becomes the current generation, and the old current generation board gets filled in as the next generation.

Another new feature here is a separate function for sending the board to *Mathematica*. Here it is called put_road. This feature is desirable whenever the code to send each generation's results is nontrivial.

Like the EpidemicC program, and all the programs to follow, this one is abortable. However, unlike EpidemicC, here there is no need to explicitly call the yield function, because it gets called automatically each generation during the MLPut calls for each road. This function is also different from EpidemicC in that we don't wait until the function finishes to begin sending data to *Mathematica*. Before the first generation begins, we call MLPutFunction to put the outer List wrapper and at that time we promise to send max_iterations + 1 roads. What do we do if we have already sent a partial answer by the time we detect an abort? We can't take back what we've sent and send Abort[] instead. It turns out that if we simply call `MLEndPacket` before we've finished sending everything we promised (*i.e.*, at a time when it is illegal), then *Mathematica* will get the symbol `$Aborted` by default. This useful behavior of *MathLink* makes it feasible to start sending data to *Mathematica* even though there's a chance we'll never finish it.

```
:Begin:
:Function:       one_lane
:Pattern:        OneLaneC[s_Integer, p_Real, vmax_Integer,
                           maxIterations_Integer]
:Arguments:      {s, p, vmax, maxIterations}
:ArgumentTypes:  {Integer, Real, Integer, Integer}
:ReturnType:     Manual
:End:
```

Here is the C code:

```c
#include "mathlink.h"
#include <stdlib.h>

#define TRUE    1
#define FALSE   0

/* Arbitrary choice outside the range of legal speeds: */
#define EMPTY_SPACE  -1

/* Increment macro that wraps around: */
#define NEXT_INDEX(i) (i != s-1 ? i+1 : 0)

#define MIN(a, b, c)
            ((tmp = ((a) < (b) ? (a) : (b))) < (c) ? tmp : (c))

void put_road(short *road, int s);
void one_lane(int s, double p, int vmax, int max_iterations);

/**************   one_lane   ***************/

void one_lane(int s, double p, int vmax, int max_iterations) {

    short   i,
            j,
            tmp,
            distance,
            num_iterations,
            speed,
            new_speed,
            *road,
            *next_road,
            *temp_ptr;

    road = (short *) calloc(s, sizeof(short));
    if(!road) {
        MLPutSymbol(stdlink, "$Failed");
        return;
    }
    next_road = (short *) calloc(s, sizeof(short));
    if(!next_road) {
        MLPutSymbol(stdlink, "$Failed");
        free(road);
        return;
    }
```

```
        /* Set up initial road configuration */

    temp_ptr = road;
    for(i = 1; i <= s; i++) {
        if( ((double) rand())/((double) RAND_MAX) < p) {
            *temp_ptr++ = (rand()>>8) % (vmax + 1);
        } else {
            *temp_ptr++ = EMPTY_SPACE;
        }
    }

    /* Send back the outer List wrapper, and the first road
       configuration */
    MLPutFunction(stdlink, "List", max_iterations + 1);
    put_road(road, s);

    for(num_iterations =1; num_iterations <= max_iterations &&
            !MLAbort; num_iterations++) {
        i = 0;
        while(i < s) {
            speed = road[i];
            if(speed == EMPTY_SPACE) {
                next_road[i] = EMPTY_SPACE;
                            /* won't necessarily stay empty */
                i++;
            } else {
                /* Count how far it is until next car */
                distance = 1;
                j = NEXT_INDEX(i);
                while(road[j] == EMPTY_SPACE) {
                    distance++;
                    j = NEXT_INDEX(j);
                }
                /* Set new speed of this car appropriately */
                new_speed = MIN(vmax, distance - 1, speed + 1);
                /* Move this car ahead that many spaces, marking
                   all the spaces we jump over as empty */
                for(j = i, tmp = 0; tmp < new_speed;
                            j = NEXT_INDEX(j), tmp++) {
                    next_road[j] = EMPTY_SPACE;
                }
                next_road[j] = new_speed;
                i += new_speed + 1;
            }
        }
        put_road(next_road, s);

        temp_ptr = road;
        road = next_road;
        next_road = temp_ptr;
    }

    /* If we finished the loop normally, we've put as many
       elements as we promised, and this MLEndPacket is
       superfluous. If we finished the loop by aborting, then
       this MLEndPacket is a deliberate error that causes
       $Aborted to be received in Mathematica. */
    MLEndPacket(stdlink);

    free(road);
    free(next_road);
}
```

```
/************* put_road *************/

void put_road(short *road, int s) {

    char  speed;
    int   i;

    MLPutFunction(stdlink, "List", s);
    for(i = 0; i < s; i++) {
        if((speed = road[i]) == EMPTY_SPACE) {
            MLPutSymbol(stdlink, "e");
        } else {
            MLPutInteger(stdlink, speed);
        }
    }
}
```

PHASES C

The phases algorithm is very straightforward to code in C. The only trick is in implementing the periodic boundary conditions. This means, for example, that the northern neighbor of a cell in the first row of the matrix is in the last row. We implement these wrap-around conditions in the NORTH, EAST, SOUTH, and WEST macros.

The important feature introduced in this simulation is the *MathLink* optimization in the put_matrix function. What we are sending to *Mathematica* each generation is a matrix of small integers. One way to do this is to put each integer separately with MLPutShortInteger. A simpler way is to use one of the PutArray functions, for example, MLPutShortIntegerArray. This will put the entire array in a single step. However, in current versions of *MathLink*, this is no faster than putting each element individually. An upcoming version will have significant optimizations, some of which will make the PutList and PutArray functions much faster. For now, though, their only advantage is convenience.

It is much faster to send the array of integers as a *Mathematica* string, then use ToCharacterCode on the string to convert it into a list of corresponding integer values. (We also need to use Partition to reconstruct the array structure.) It is very quick to send a block of data as a single string, and it so happens that ToCharacterCode is an extremely fast function. This technique of sending a string wrapped in ToCharacterCode is by far the fastest way to send a list of small integers.

You'll note we referred above to sending a *Mathematica* string, which is different from a C string. *Mathematica* strings can contain any eight-bit character value, including 0, which cannot appear appear in a C string because it is used to mark the end of the string. Since *MathLink* is C-based, it needs to have a special representation for *Mathematica* strings. If we try to use `MLPutString` to send an array of data, *MathLink* will stop at the first 0 byte it encounters. We need to send zeros as part of our data in this simulation, so we need to convert each `cell_matrix` into a legal C representation of a *Mathematica* string before sending it with `MLPutString`. This transformation happens in the `put_cell_matrix` function.

The main function is `MLPutCharToString`. This takes a `char` and converts it into the appropriate encoding in a *Mathematica* string. All the characters with ASCII values between 1 and 127 are left unchanged by this function, but 0 and ASCII 128–255 are encoded in a special way that generally occupies two characters. Thus, the *Mathematica* string will generally be longer than the C string, so the first step is to figure out how much memory we need to hold the new string. If we call `MLPutCharToString` with `NULL` as the second argument (where the destination memory location normally goes), it performs no action, but simply returns the number of characters required to encode the given character. Therefore, we simply call `MLPutCharToString` on every character in the original string, accumulating the necessary size as we go. Then, we allocate that much memory and march through the string a second time calling `MLPutCharToString`, only this time we actually write the encoded string into the destination memory. The final string is then given a 0 terminating byte and sent with `MLPutString`. It may seem inefficient to have to walk through the data twice, but this step is very quick.

If we were writing this function from scratch (rather than duplicating a version already written in *Mathematica*), we could avoid the use of `MLPutCharToString` by simply choosing a different encoding of the states. As long as we stick to the integers between 1 and 127, our `cell_matrix` will be a valid C string (once we give it a terminating 0 byte) and we can simply use `MLPutString` to send it. We want to emphasize, though, that the step of converting a block of character data into a *Mathematica* string is very fast.

Here is what the template looks like:

```
:Begin:
:Function:      phases
:Pattern:       PhasesC[s_Integer, n_Integer, maxIterations_Integer]
:Arguments:     {s, n, maxIterations}
:ArgumentTypes: {Integer, Integer, Integer}
:ReturnType:    Manual
:End:
```

The C code follows:

```c
#include "mathlink.h"
#include <stdlib.h>

#define TRUE 1
#define FALSE 0

#define NORTH(i)   ((i) == 1 ? s_sqr - s : -s)
#define EAST(i)    ((i) == s ? 1 - s     : 1 )
#define SOUTH(i)   ((i) == s ? s - s_sqr : s )
#define WEST(i)    ((i) == 1 ? s - 1     : -1)

void phases(int s, int n, int max_iterations);
int put_cell_matrix(char *matrix, int s);

/************* phases **************/

void phases(int s, int n, int max_iterations) {

    int    i,
           j,
           index,
           s_sqr,
           num_iterations;
    char   state,
           *cell_matrix,
           *next_cell_matrix,
           *temp_ptr;

    s_sqr = s * s;

    /* Use calloc so that matrix starts out filled with 0's */
    cell_matrix = (char*) calloc(s_sqr, sizeof(char));
    if(!cell_matrix) {
        MLPutSymbol(stdlink, "$Failed");
        return;
    }
    next_cell_matrix = (char*) calloc(s_sqr, sizeof(char));
    if(!next_cell_matrix) {
        MLPutSymbol(stdlink, "$Failed");
        free(cell_matrix);
        return;
    }

    /* Set up the initial random configuration */
    temp_ptr = cell_matrix;
    for(i = 1; i <= s_sqr; i++) {
        *temp_ptr++ = (char) ((rand()>>8) % n);
    }

    MLPutFunction(stdlink, "List", max_iterations+1);
    if(!put_cell_matrix(cell_matrix, s)) MLAbort = TRUE;
```

```
        for(num_iterations = 1; num_iterations <= max_iterations &&
                    !MLAbort; num_iterations++) {
            index = 0;
            for(i = 1; i <= s; i++) {
                for(j = 1; j <= s; j++, index++) {
                    state = cell_matrix[index];
                    if(state == n-1) {
                        if(!(cell_matrix[index + NORTH(i)] &&
                             cell_matrix[index + EAST(j)] &&
                             cell_matrix[index + SOUTH(i)] &&
                             cell_matrix[index + WEST(j)])) {
                            next_cell_matrix[index] = 0;
                        } else {
                            next_cell_matrix[index] = state;
                        }
                    } else {
                        if(cell_matrix[index + NORTH(i)] == state+1 ||
                           cell_matrix[index + EAST(j)]  == state+1 ||
                           cell_matrix[index + SOUTH(i)] == state+1 ||
                           cell_matrix[index + WEST(j)]  == state+1) {
                            next_cell_matrix[index] = state + 1;
                        } else {
                            next_cell_matrix[index] = state;
                        }
                    }
                }
            }
            if(!put_cell_matrix(next_cell_matrix, s)) MLAbort = TRUE;

            temp_ptr = cell_matrix;
            cell_matrix = next_cell_matrix;
            next_cell_matrix = temp_ptr;
        }

        MLEndPacket(stdlink);

        free(cell_matrix);
        free(next_cell_matrix);
}

/************ put_cell_matrix *************/
int put_cell_matrix(char *cell_matrix, int s) {

    size_t i,
           size,
           matrix_len = s * s;
    char   *str, *p;

    for(i = 0, size = 0; i < matrix_len; i++)
        size += MLPutCharToString(cell_matrix[i], NULL);
    p = str = (char *) malloc(size + 1);
    if(p) {
        for(i = 0; i < matrix_len; i++)
            MLPutCharToString(cell_matrix[i], &p);
        *p = '\0';
        MLPutFunction(stdlink, "Partition", 2);
        MLPutFunction(stdlink, "ToCharacterCode", 1);
        MLPutString(stdlink, str);
        MLPutInteger(stdlink, s);
        free(str);
        return(TRUE);
    } else {
        return(FALSE);
    }
}
```

SANDPILEC

The `SandpileC` simulation is very much like `PhasesC`, but with one important new feature: We don't know ahead of time how many generations the simulation will run. This program will reach a point where the board doesn't change from one generation to the next, and we want to stop immediately when this happens. You'll notice in the *Mathematica* version that `FixedPointList` is used to stop automatically when the board no longer changes, but of course in C we are responsible for coding the logic to detect this situation. The logic is quite simple; we start each generation on the assumption that it will be the last, and then cancel this assumption any time we make a change to the board.

Whereas it is simple to implement the stopping algorithm, it is more complicated to deal with the *MathLink* ramifications. In the previous simulations, we used the technique of sending each generation's results as they are generated, and we want to continue this. This requires that we send the outer `List` wrapper that will enclose all the single-generation results before we even begin the generation loop. When we send this enclosing `List`, we need to specify the number of elements that will be in it, and therein lies the problem. At the time we send the outer `List`, the only thing we know is the maximum requested number of generations. If we promise to send this maximum number, how do we fulfill this promise if the simulation runs to completion in a smaller number of generations?

There are several methods for dealing with this issue, and as usual we choose a simple one. We rely on the properties of a special *Mathematica* function called `Sequence`. `Sequence` has the property that it disappears whenever it is inside another expression. In other words, `f[a, Sequence[b, c], d]` becomes `f[a, b, c, d]`. In effect, `Sequence` "dissolves" whenever another function is wrapped around it. Another example of the behavior of `Sequence` is that `{a, b, c, Sequence[], Sequence[]}` evaluates to `{a, b, c}`. That is, `Sequence` with no arguments simply vanishes altogether. You can probably see where this is going. If we promise 20 generations in the result `List`, but the simulation ends after 10, we can simply send `Sequence[]` objects to fill out the remaining 10 elements. When the resulting list is evaluated, the `Sequence[]` placeholders will disappear, leaving a list of 10 matrices.

Another difference between this and the `PhasesC` code is that here we have a border of zeros that surround the inner matrix of cells. We never have to update the cells in this inert border, so we don't have to worry about stepping outside the boundaries of the array when looking at neighbor sites. This introduces a bit of complexity when incrementing a pointer to walk through the array, since we need to increment two extra times when crossing from the end of one row to the beginning of the next.

Like `PhasesC`, we use the technique of sending each generation's matrix as a string wrapped in `ToCharacterCode` and `Partition`. We have to use `MLPutCharToString` to encode the strings, since they contain zeros (from the border only). If we weren't trying to duplicate the *Mathematica* version, we would probably choose a different representation of the border to avoid this step.

```
:Begin:
:Function:       sandpile
:Pattern:        SandpileC[s_Integer, m_Integer]
:Arguments:      {s, m}
:ArgumentTypes:  {Integer, Integer}
:ReturnType:     Manual
:End:

#include "mathlink.h"
#include <stdlib.h>

#define TRUE    1
#define FALSE   0

void sandpile(int s, int max_iterations);
int put_landscape(char *landscape, int dim);
short count_topheavy_neighbors(int i, int j);

char    *landscape;  /* convenient to have these globals */
size_t  dim;

/**************  sandpile  **************/

void sandpile(int s, int max_iterations) {

    int     i,
            j,
            randx,
            randy,
            done,
            num_iterations;
    char    height,
            *next_landscape,
            *ptr1,
            *ptr2,
            *temp;

    dim = s + 2;
    landscape = (char *) calloc(dim * dim, sizeof(char));
    if(!landscape) {
        MLPutSymbol(stdlink, "$Failed");
        return;
    }
    next_landscape = (char *) calloc(dim * dim, sizeof(char));
    if(!next_landscape) {
        MLPutSymbol(stdlink, "$Failed");
        free(landscape);
        return;
    }
```

```
/* Note the pointer increment tricks used here to walk through
   the array, skipping over the border of zeros that surround
   the true board */
ptr1 = landscape + dim + 1;
for(i = 1; i <= s; i++, ptr1 += 2) {
    for(j = 1; j <= s; j++, ptr1++) {
        *ptr1 = (char) ((rand()>>8) % 2 + 3);
    }
}

/* Set up the initial values */
do {
    randx = (rand()>>4) % s + 1;
    randy = (rand()>>4) % s + 1;
} while(++landscape[randx * dim + randy] < 5);

/* Promise to send the maximum requested number of generations,
   even though the simulation will likely end before this number
   is reached */
MLPutFunction(stdlink, "List", max_iterations+1);
if(!put_landscape(landscape, dim)) MLAbort = TRUE;

num_iterations = 0;
do {
    done = TRUE;
    num_iterations++;
    ptr1 = landscape + dim + 1;
    ptr2 = next_landscape + dim + 1;
    for(i = 1; i <= s; i++, ptr1 += 2, ptr2 += 2) {
        for(j = 1; j <= s; j++, ptr1++, ptr2++) {
            height = *ptr1;
            if(height <= 4) {
                *ptr2 = height +
                        count_topheavy_neighbors(i, j);
            } else {
                done = FALSE;
                *ptr2 = height +
                        count_topheavy_neighbors(i, j) - 4;
            }
        }
    }

    if(!put_landscape(next_landscape, dim)) MLAbort = TRUE;

    temp = landscape;
    landscape = next_landscape;
    next_landscape = temp;

} while(!done && num_iterations < max_iterations && !MLAbort);

if(!MLAbort) {
    /* In case the simulation ended before we got to
       max_iterations, fill out the promised length of the
       result list by sending Sequence[] objects, which will
       dissolve in place when the list hits the evaluator */
    for(i = num_iterations + 1; i <= max_iterations; i++)
        MLPutFunction(stdlink, "Sequence", 0);
}
MLEndPacket(stdlink);

free(landscape);
free(next_landscape);
}
```

```
/********   count_topheavy_neighbors  **********/

#define TOPHEAVYQ(x) ((x) < 5 ? 0 : 1)

/* mult could be eliminated by passing in not i, j,
   but ptr1-landscape */
short count_topheavy_neighbors(int i, int j) {
    size_t index = i * dim + j;
    return TOPHEAVYQ(landscape[index - dim]) +
           TOPHEAVYQ(landscape[index + 1]) +
           TOPHEAVYQ(landscape[index - 1]) +
           TOPHEAVYQ(landscape[index + dim]);
}

/*************   put_landscape   *************/

int put_landscape(char *landscape, int dim) {

    size_t i,
           size,
           landscape_len = dim * dim;
    char   *str, *p;

    /* This method is discussed in conjunction with the
       PhasesC program */
    for(i = 0, size = 0; i < landscape_len; i++)
        size += MLPutCharToString(landscape[i], NULL);
    p = str = (char *) malloc(size + 1);
    if(p) {
        for(i = 0; i < landscape_len; i++)
            MLPutCharToString(landscape[i], &p);
        *p = '\0';
        MLPutFunction(stdlink, "Partition", 2);
        MLPutFunction(stdlink, "ToCharacterCode", 1);
        MLPutString(stdlink, str);
        MLPutInteger(stdlink, dim);
        free(str);
        return(TRUE);
    } else {
        return(FALSE);
    }
}
```

LifeGameC

The `LifeGameC` program incorporates all of the *MathLink* techniques introduced so far, and introduces a couple useful tricks in the template entry itself. We want the `LifeGameC` program to be able to accept an initial configuration specification, but default to a random one if none is supplied. If we allow the user to specify a list of initial cells to be occupied, there is the possibility that they will include some sites that are outside the boundaries of the board. We need to trap this error at some point, since the program will surely crash if we try to write outside the array.

There are many ways to catch this error, but it is most convenient to do it in *Mathematica* itself, before the external function is actually called. Let's say we want the function to detect an improper initial configuration, issue an error message, and then default to using a randomly generated one. Earlier we noted that you can put arbitrary *Mathematica* code on the :Arguments: line of a template entry. Whatever this code evaluates to is what will be sent to the external program. Therefore, we can just put the code that tests the initial configuration right on the :Arguments: line. If the initial configuration is valid, we pass it to the external program (after flattening for convenience); if not, we issue a message and pass an empty list, which the program interprets as a request to create a random configuration.

We want to issue a message called LifeGameC::input, but how do we get the definition of this message into *Mathematica*? We use another feature of template entries, the :Evaluate: line. Any *Mathematica* code that appears on an :Evaluate: line is sent to *Mathematica* when the external program is installed. Thus, we can send along any auxiliary code we need to support the template functions, messages being a good example. Here is what the template file looks like:

```
:Evaluate:        LifeGameC::input = "Initial input configuration
                                      out of range. Defaulting to
                                      random configuration."

:Begin:
:Function:        life
:Pattern:         LifeGameC[s_Integer, maxGens_Integer,
                      initial_List:{}]
:Arguments:       {s, maxGens, If[Max[initial] <= s &&
                                Min[initial] >= 1,
                                Flatten[initial],
                                Message[LifeGameC::input]; {}] }
:ArgumentTypes:   {Integer, Integer, IntegerList}
:ReturnType:      Manual
:End:
```

You will notice that one of the arguments passed to the life function is an integer list, indicated by the keyword IntegerList on the :ArgumentTypes: line. Whenever we use IntegerList or RealList as one of the argument types, our function needs to be written to take a long parameter immediately following the list itself. The code that **mprep** writes to read the input list will determine the length of the list and pass it to our function. Thus, the prototype for the life function is as follows:

```
void life(int s, int max_gens, int initial[], long len);
```

The game of Life has been coded in C countless times. Programmers have tried to squeeze it into the smallest possible space and optimize it down to the last millisecond. Our approach makes no attempt to compete on these terms, and thus ends up being simple and short, although still quite fast.

```
#include "mathlink.h"
#include <stdlib.h>

#define TRUE    1
#define FALSE   0

#define ALIVE   1
#define DEAD    0

#define NORTH(i)             ((i) == 1 ? s_sqr - s : -s)
#define EAST(j)              ((j) == s ? 1 - s      : 1 )
#define SOUTH(i)             ((i) == s ? s - s_sqr  : s )
#define WEST(j)              ((j) == 1 ? s - 1      : -1)
#define NORTHEAST(i, j)   NORTH(i) + EAST(j)
#define SOUTHEAST(i, j)   SOUTH(i) + EAST(j)
#define SOUTHWEST(i, j)   SOUTH(i) + WEST(j)
#define NORTHWEST(i, j)   NORTH(i) + WEST(j)

void life(int s, int max_gens, int initial[], long len);
int put_board(char *board, int s);

/**************** life ****************/

void life(int s, int max_gens, int initial[], long len) {

    long    i,
            j,
            s_sqr;
    int     num_gens,
            done,
            index,
            x_index,
            y_index,
            num_living_neighbors;
    char    state,
            *board,
            *next_board,
            *temp_ptr;

    s_sqr = s * s;
    /* calloc is used to fill the arrays with 0's initially,
       corresponding to all DEAD sites */
    board = (char*) calloc(s_sqr, sizeof(char));
    if(!board) {
        MLPutSymbol(stdlink, "$Failed");
        return;
    }
    next_board = (char*) calloc(s_sqr, sizeof(char));
    if(!next_board) {
        MLPutSymbol(stdlink, "$Failed");
        free(board);
        return;
    }

    /* len is the length of the initial configuration supplied
       by the user. If it is 0, none was supplied, so we default
       to a random one. Otherwise, we set the board up according
       to the specification */
```

```
    if(len == 0) {
        temp_ptr = board;
        for(i = 0; i < s_sqr; i++, temp_ptr++)
            if((rand()>>8) % 2)
                *temp_ptr = ALIVE;
    } else {
        for(i = 0; i < len/2; i++) {
            x_index = *initial++;
            y_index = *initial++;
            board[(y_index - 1) * s + (x_index - 1)] = ALIVE;
        }
    }

    MLPutFunction(stdlink, "List", max_gens + 1);
    if(!put_board(board, s)) MLAbort = TRUE;

    num_gens = 0;
    do {
        done = TRUE;
        num_gens++;
        index = 0;
        for(i = 1; i <= s; i++) {
            for(j = 1; j <= s; j++, index++) {
                state = board[index];
                num_living_neighbors =
                        (board[index + NORTH(i)] == ALIVE) +
                        (board[index + NORTHEAST(i, j)] == ALIVE) +
                        (board[index + EAST(j)] == ALIVE) +
                        (board[index + SOUTHEAST(i, j)] == ALIVE) +
                        (board[index + SOUTH(i)] == ALIVE) +
                        (board[index + SOUTHWEST(i, j)] == ALIVE) +
                        (board[index + WEST(j)] == ALIVE) +
                        (board[index + NORTHWEST(i, j)] == ALIVE);
                if(state == ALIVE) {
                    if(num_living_neighbors == 2 ||
                       num_living_neighbors == 3) {
                        next_board[index] = ALIVE;
                    } else {
                        next_board[index] = DEAD;
                        done = FALSE;
                    }
                } else {    /* current state is DEAD */
                    if(num_living_neighbors == 3) {
                        next_board[index] = ALIVE;
                        done = FALSE;
                    } else {
                        next_board[index] = DEAD;
                    }
                }
            }
        }

        if(MLAbort) {
            done = TRUE;
        } else {
            if(!put_board(next_board, s)) {
                MLAbort = TRUE;
                done = TRUE;
            }
        }
        temp_ptr = board;
        board = next_board;
        next_board = temp_ptr;
    } while(!done && num_gens < max_gens);
```

```
    if(!MLAbort) {
        /* In case the simulation ended before we got to max_gens,
           fill out the promised length of the result list by
           sending Sequence[] objects, which will dissolve in
           place when the list hits the evaluator */
        for(i = num_gens + 1; i <= max_gens; i++)
            MLPutFunction(stdlink, "Sequence", 0);
    }
    MLEndPacket(stdlink);

    free(board);
    free(next_board);
}

/************** put_board ***************/

int put_board(char *board, int s) {

    size_t i,
           size,
           board_len = s * s;
    char   *str, *p;

    for(i = 0, size = 0; i < board_len; i++)
        size += MLPutCharToString(board[i], NULL);
    p = str = (char *) malloc(size + 1);
    if(p) {
        for(i = 0; i < board_len; i++)
            MLPutCharToString(board[i], &p);
        *p = '\0';
        MLPutFunction(stdlink, "Partition", 2);
        MLPutFunction(stdlink, "ToCharacterCode", 1);
        MLPutString(stdlink, str);
        MLPutInteger(stdlink, s);
        free(str);
        return(TRUE);
    } else {
        return(FALSE);
    }
}
```

Appendix D
Remote Computing with Mathematica

INTRODUCTION

For the simulations in this book, you will usually find that your computer works fine most of the time. Occasionally, you will want to work on a larger lattice, or let a simulation run through many more generations, and you might find yourself waiting for longer than you would like. This is especially critical if you are looking for patterns in a model and need to run a simulation over and over until something interesting crops up. When this happens, you will find yourself looking for ways to avoid computational bottlenecks. This appendix describes how to link your computer to other machines that either have more "horsepower" or are not as busy as yours. This will allow you to work locally on your computer while computations are performed on another, remote machine.

There are two reasons why you might want to run your *Mathematica* computations on a remote machine. The most common reason is that your computer may not be fast enough or contain enough memory to run a large-scale problem. Running 1000 generations of a simulation on a 256×256 lattice with 14 values at each site is not meant for a machine with a slow processor and 8 MB of memory.

The other reason that you would want to compute remotely relates only to multitasking systems such as Unix. On these systems, dozens or even hundreds of users can be logged in simultaneously, each using valuable resources and CPU time. This can slow your computations down to a crawl under certain circumstances. Some Unix users try to avoid this situation by running their computations at night when usage should be lower, but this can also pose problems, especially if others have similar ideas.

Other than *MathLink*, which is discussed in Appendix C, there are two ways to compute remotely. You can either use the kernel on a remote machine that is connected to your network while using the front end on your local machine, or you can do the computations on a remote machine and then transfer the results back to your local site. The next two sections will describe each of these processes in some detail. For more complete descriptions and, most importantly, for machine-specific instructions, you should consult the User's Guide that comes with your copy of *Mathematica*.

D.1 ■ REMOTE COMPUTING WITH A LOCAL FRONT END

You can run the *Mathematica* front end (the graphical interface) on your local machine while expressions are evaluated on a different (remote) machine. The remote machine must be connected to your local computer either by means of a serial device (modem or cable) or network (LocalTalk or Ethernet). When running *Mathematica* in this manner, you will not notice anything different except for the fact that your computations will run at the speed of the remote machine. Everything else will look and act the same. So, for example, you can edit and save files using your local front end as you did before.

In a typical scenario, you may wish to connect to a Unix mainframe running *Mathematica* that can be reached on your local area network. For other scenarios (such as dialing-in to a remote kernel and using remote kernels intra-network), the details of setting up are different, but using such a system is the same as what is described here. Consult your User's Guide for the configuration instructions in these cases.

To get started you will first have to determine whether your computer is connected to other machines and whether those machines have *Mathematica* installed on them. If you are not sure, you will have to ask your system administrator for this information. You will need to make note of the "hostname" of the computer that will be running your remote kernel.

Once you have determined that a connection exists, start up *Mathematica* on your computer (we'll refer to this machine as "local"). Then, in the **Action** menu, select **Kernels and Tasks**. A dialog box will appear to allow you to set up the remote connection.

Follow the instructions in your User's Guide in the chapter "Interacting with the Kernel" and fill in the entries in this dialog box and any others in the Connections Settings and Terminal Settings dialog boxes. Most of the settings you select in the Connections and Terminals dialog boxes are machine and network specific, so it is impossible to give details on this here. If you are not sure of any of these, consult your system administrator. Typically, the name of the kernel program on a Unix machine is **math**. You can give any name you like to the Connection name. Type in the hostname of the remote computer that the kernel will be run on in the appropriate box.

One point to be aware of concerns security. During the set-up of the remote kernel, you will have the opportunity to enter login strings to automate the task of connecting to the remote kernel. Typically, you would give your username and password in the dialog box and then these strings would be used to connect to the remote computer. If you are the only one with access to your local computer, this is fine, but in a lab setting, it is best to leave these items blank. In this case, you will be prompted to enter a username and password each time you start the remote kernel.

Once you have completed the set-up procedure (or if it has already be done for you), you are ready to make the connection. Select **Kernels and Tasks** from the **Action** menu, select the name of your remote kernel, and click Launch Kernel. (On a local Macintosh, you would select **Current** from the **Kernel** menu and then select the remote kernel.) Type some input (2 + 3, for example) in a notebook on your local machine and evaluate. If you had selected "Append name to In/Out prompts" during the connection set-up, then you would see this reflected on screen. Below are two typical examples.

```
(Remote kernel) In[1]:= 2 + 3

(Remote kernel) Out[1]= 5


(HP 9000) In[1]:= 100!

(HP 9000) Out[1]= 2432902008176640000
```

Once you are satisfied that your connection is established and working properly, you might want to try some tests to get a sense of the kinds of increases in computational power to expect. For example, the authors have used the following two tests to get rough estimates of relative speeds on different platforms. The first computation uses integer (machine and high-precision) arithmetic, while the computation of the determinant of a dense 1000×1000 matrix gives a good measure of floating point capabilities.

```
(DEC OSF/1 Alpha) In[1]:= Timing[ 10000!; ]

(DEC OSF/1 Alpha) Out[1]= {3.28333 Second, Null}


(DEC OSF/1 Alpha) In[2]:= Timing[ Det[Table[Random[],
                                      {1000}, {1000}]]; ]

(DEC OSF/1 Alpha) Out[2]= {142.017 Second, Null}
```

For comparison, here are the corresponding timings on a 68030 desktop computer (circa 1994).

```
(Macintosh 030) In[1]:= Timing[ 10000!; ]

(Macintosh 030) Out[1]= {91.7833 Second, Null}


(Macintosh 030) In[2]:= Timing[ Det[Table[Random[],
                                   {1000}, {1000}]]; ]

(Macintosh 030) Out[2]= {12966.15 Second, Null}
```

(Caveat: These computations are not meant as a scientific analysis of relative performance and should be taken with a grain of salt. Although they give a good "rough" guide to the kinds of differences on different platforms, they should not be used in an absolute sense. Factors such as different hardware and software configurations and user usage on Unix machines will affect such results.)

Besides the fact that great improvements in speed are possible by using a powerful remote kernel, some computations are just impossible to carry out on some desktop computers. If your computer does not have enough memory to perform a lengthy computation (nor disk space to store the result), its relative speed is immaterial.

D.2 ■ REMOTE COMPUTING ACROSS NETWORKS

During the writing of this book, one of the authors did not have direct access to a network for a period of time and had to run many of the simulations on a computer 2000 miles away. Fortunately, he could dial in to a remote network via modem. Not knowing how long some simulations would run and wanting to keep his phone bills to a minimum, he simply sent his input files to the remote machine, instructed it to carry out the computation and to save the result in a file on the remote computer. He then logged out and checked the computation every few days to see when it was done. Once it had terminated, he transferred the results back to his local site. This section describes the process to follow to compute *across* remote networks.

Assuming you can make a connection to a remote machine running a *Mathematica* kernel (here assumed to be a Unix computer) via such protocols as FTP and telnet, the process is quite straightforward: Prepare an input file locally, transfer the input file to the remote machine, perform the computation, and finally retrieve the results by transferring any output back to your local machine.

To demonstrate this process we will create a short file, `walk3d.m`, that defines the function `Walk3D` and performs a computation using this function—a 10,000-step random walk on a three-dimensional lattice. Here are the entire contents of the input file:

```
Walk3D[n_] :=
   FoldList[Plus, {0, 0, 0},
         {{-1,1,1}, {-1,-1,1}, {1,-1,1},
          {1,1,1}, {-1,1,-1}, {-1,-1,-1},
          {1,-1,-1}, {1,1,-1}}[[Table[
               Random[Integer, {1, 8}], {n}]]]
         ]

Walk3D[10000]
```

The details of the file transfer will depend upon your particular set-up. Typically, you will use FTP (file transfer protocol) or Kermit, or ZModem, although for small files, you could use telnet and copy and paste the contents of your file into an editor. You can transfer the file as a text file (ASCII) or archive and/or compress the file using utilities. Make sure the same utilities are available on the remote computer so the file can be restored to its original state.

Once the file is transferred, connect to the remote computer (typically with telnet, or by dialing in). Log in with your username and password and check that *Mathematica* can be accessed by typing **math** to start up the kernel. In the following example, the name of the remote computer is "pyrethrum," shown as part of the Unix prompt here.

```
pyrethrum [1] wellin> math
Using distribution image.
Mathematica 2.2 for DEC OSF/1 Alpha
Copyright 1988-94 Wolfram Research, Inc.
 - Terminal graphics initialized -

In[1]:=
```

You can check that the file `walk3d.m` is in the current working directory of the remote machine:

```
In[1]:= FileNames[]

Out[1]= {AppendixD.ma, data, dla1000.data, dla.m,
          PartialDifferentialOperators.ma,
          walk3d.m}
```

The file can now be loaded in the current session on the remote computer. Since the output from this file will be 10,000 points in 3-space, it will be quite long. Instead of sending it to the screen, we will direct it to the file `walk3d.data`.

```
In[2]:= << walk3d.m >> walk3d.data
```

A quick check shows the (abbreviated) contents of the file:

```
In[3]:= Short[ <<walk3d.data ]

Out[3]//Short= {{0, 0, 0}, {-1, 1, -1}, <<9998>>, {-142, 88, 24}}
```

To bring the results back to your local computer, again use your file transfer utilities (FTP, Kermit, ZModem, etc.), possibly compressing the file before retrieving it. If the file is compressed, remember to perform the transfer in binary mode.

Although the above computation should take no more than a minute or two on most workstations, some of the simulations in this book require considerably more time to run. In fact, some of the simulations took over a week to run on some of today's fastest workstations! Most communication packages will automatically terminate your session ("time you out") if there is no keyboard activity for 30 minutes or so. As opposed to pressing the space bar on your keyboard every 29 minutes, you can use the Unix utility **nohup** to prevent the process from being interrupted. The following input causes the execution of command and does not allow it to be interrupted due to hanging up or logging out.

```
prompt> nohup command
```

So, using the file `walk3d.m` defined above, you can start a process that cannot be interrupted by logging out as follows:

```
prompt> nohup math -batchinput -batchoutput\
        < walk3d.m > walk3d.data &
```

This starts up *Mathematica* (math) allowing no interruption (nohup), taking commands (-batchinput) from the input file `walk3d.m`, and writing output (-batchoutput) to the file `walk3d.data`. Placing the & at the end of the line is the standard way to make sure the job runs in the background, giving you the shell prompt back again, so that you can continue to do other work or log out without waiting for the process to terminate.

You will periodically want to check the state of your computation. You can include `Print` statements in the program to indicate what stage the program has just completed. This can be checked by logging in to the remote computer and checking the output file with a text editor.

Another way to check the status of your computation is to use the Unix **ps** command:

```
prompt> ps -ax | grep math
```

This will show all running processes involving the `math` command. If you don't see your process there, it is done.

For a more complete discussion of **nohup** and related Unix commands, see the references given below.

REFERENCES

Steven M. Christensen. History, nohup, and the end.m. *The Mathematica Journal* 3 (Spring 1993) 57–59.

Todd Gayley. Tech Support: Remote kernels, ComplexExpand, and memory use under Windows. *The Mathematica Journal* 4 (1994) 70–73.

Roman E. Maeder. *Programming in Mathematica*, Second Edition. Addison-Wesley. 1991.

APPENDIX E

MathLink Program Listing

by Todd Gayley

INTRODUCTION

This appendix contains the full source code for the *MathLink* functions that are described in Appendix C of this book. Each source code is given as a .tm file, meaning the *MathLink* template is included with the C source code.

E.1 THE RANDOM WALK

```
/******************************************************************

                         walk2D.tm

    **************************************************************/


/*****************  Templates begin  *****************/

:Begin:
:Function:       walk2D
:Pattern:        Walk2DC[n_Integer]
:Arguments:      {n}
:ArgumentTypes:  {Integer}
:ReturnType:     Manual
:End:
```

```
:Begin:
:Function:      walk2D_array
:Pattern:       Walk2DCArray[n_Integer]
:Arguments:     {n}
:ArgumentTypes: {Integer}
:ReturnType:    Manual
:End:

:Begin:
:Function:      seed_random
:Pattern:       SeedRandomC[seed_Integer]
:Arguments:     {seed}
:ArgumentTypes: {Integer}
:ReturnType:    Manual
:End:


/****************** C code begins ******************/

#include "mathlink.h"
#include <stdlib.h>

void walk2D_array(int n);
void walk2D(int n);
void seed_random(int seed);


/****************** walk2D ******************/

void walk2D(int n) {

    int i,
        x,
        y;

    x = y = 0;  /* Starting values of x and y positions */

    /* We put the head List right away */
    MLPutFunction(stdlink, "List", n + 1);

    /* Each element of the result list is itself a list of x
       and y values. Here we put the first element, {0, 0} */
    MLPutFunction(stdlink, "List", 2);
    MLPutInteger(stdlink, x);
    MLPutInteger(stdlink, y);
```

```
    for(i = 1; i <= n; i++) {
        switch((rand()>>10) % 4) {
            case 0:  x++; break;
            case 1:  y++; break;
            case 2:  x-; break;
            case 3:  y-;
        }
        MLPutFunction(stdlink, "List", 2);
        MLPutInteger(stdlink, x);
        MLPutInteger(stdlink, y);
    }
}


/******************* walk2D_array *******************/

void walk2D_array(int n) {

    int    i;
    short  x,
           y,
           *index,
           *array;
    long   dims[2];

    dims[0] = n + 1;
    dims[1] = 2;

    /* Here we allocate memory to hold the entire array */
    array = malloc((n + 1) * 2 * sizeof(short));
    /* The memory allocation might fail, and if so we want to bail out
       of the function immediately. */
    if(!array) {
        MLPutSymbol(stdlink, "$Failed");
        return;
    }

    x = y = array[0] = array[1] = 0;
    index = array + 2;  /* for the first x=0, y=0 */
    for(i = 1; i <= n; i++) {
        /* index is the pointer into the array. A statement like
           *index++ = x means "put the value in x at the address pointed
           to by index, then increment index" (so it points to the next
           element in the array) */
        switch((rand()>>10) % 4) {
            case 0:  *index++ = ++x;
                     *index++ = y;
                     break;
```

```
                case 1:   *index++ = x;
                          *index++ = ++y;
                          break;
                case 2:   *index++ = -x;
                          *index++ = y;
                          break;
                case 3:   *index++ = x;
                          *index++ = -y;
            }
        }
    MLPutShortIntegerArray(stdlink, array, dims, NULL, 2);
    free(array);
}


/*************  seed_random  *************/

void seed_random(int seed) {

    srand((unsigned int) seed);
    MLPutSymbol(stdlink, "Null");
}


/**************  main  ***************/

#if !WINDOWS_MATHLINK
int main(int argc, char *argv[]) {
    return MLMain(argc, argv);
}
#else
int PASCAL WinMain(HANDLE hinstCurrent, HANDLE hinstPrevious,
                   LPSTR lpszCmdLine, int nCmdShow) {

    char  buff[512];
    char  FAR * argv[32];
    int   argc;

    if(!MLInitializeIcon(hinstCurrent, nCmdShow)) return 1;
    argc = MLStringToArgv(lpszCmdLine, buff, argv, 32);
    return MLMain(argc, argv);
}
#endif
```

E.2 ■ EPIDEMICS

```
/*****************************************************************


                        epidemic.tm


*****************************************************************/


/***************** Templates begin ******************/

:Begin:
:Function:        epidemic
:Pattern:         EpidemicsC[max_Integer, p_Real]
:Arguments:       {max, p}
:ArgumentTypes:   {Integer, Real}
:ReturnType:      Manual
:End:

:Begin:
:Function:        seed_random
:Pattern:         SeedRandomC[seed_Integer]
:Arguments:       {seed}
:ArgumentTypes:   {Integer}
:ReturnType:      Manual
:End:

/***************** C code begins ******************/

#include "mathlink.h"
#include <stdlib.h>

#define TRUE 1
#define FALSE 0

/* Note the hard-coded limit. This is to avoid memory allocations of
   >64 Kb in Windows. If using Macintosh or Unix (or a Win32 program
   under Windows), you can increase this number as much as you like. */
#define MAX_SITES  20000
#define CLUSTER       1
#define PERIMETER     2
#define REJECT        3

struct site {
            char xcoord;
            char ycoord;
            char type;
};
```

```
void epidemic(int max_size, double p);
void seed_random(int seed);

/******************  epidemic  ******************/

void epidemic(int max_size, double p) {

    int     i,
            perimeter_length,
            cluster_length,
            sites_length,
            chosen_site,
            north_empty,
            south_empty,
            east_empty,
            west_empty;
    char    x,
            y,
            y_north,
            y_south,
            x_east,
            x_west;
    struct site  *sites,
                 *site_ptr;

    sites = (struct site *) malloc(MAX_SITES * sizeof(struct site));
    if(!sites) {
        MLPutSymbol(stdlink, "$Failed");
        return;
    }

    sites[0].xcoord = 0;
    sites[0].ycoord = 0;
    sites[0].type = CLUSTER;
    sites[1].xcoord = 1;
    sites[1].ycoord = 0;
    sites[1].type = PERIMETER;
    sites[2].xcoord = 0;
    sites[2].ycoord = 1;
    sites[2].type = PERIMETER;
    sites[3].xcoord = -1;
    sites[3].ycoord = 0;
    sites[3].type = PERIMETER;
    sites[4].xcoord = 0;
    sites[4].ycoord = -1;
    sites[4].type = PERIMETER;
```

```
cluster_length = 1;
perimeter_length = 4;
sites_length = 5;

while(perimeter_length > 0 &&
        cluster_length < max_size &&
            sites_length <= MAX_SITES - 3 &&
                !MLAbort) {

    /* Choose a random perimeter site */
    chosen_site = rand() % perimeter_length + 1;
    /* For efficiency, we start at end of sites array and count
      backward, since perimeter sites are mainly clustered at the
        end of the array */
    site_ptr = sites + sites_length;
    for(i = 1; i <= chosen_site; i++) {
        do site_ptr-; while(site_ptr->type != PERIMETER);
    }
    if((1.0 * rand())/RAND_MAX > p) {
        site_ptr->type = REJECT;
        perimeter_length-;
    } else {
        site_ptr->type = CLUSTER;
        cluster_length++;
        perimeter_length-;
        x = site_ptr->xcoord;
        y = site_ptr->ycoord;
        y_north = y + 1;
        y_south = y - 1;
        x_east = x + 1;
        x_west = x - 1;
        north_empty = south_empty = east_empty = west_empty = TRUE;

        /* We make one pass through the array looking for non-empty
            neighbor sites; otherwise, neighbor sites are added as
                new perimeter sites. */
        for(site_ptr = sites, i = 0; i < sites_length;
                i++, site_ptr++) {
            char temp_x = site_ptr->xcoord,
                 temp_y = site_ptr->ycoord;
            if(temp_x == x)
                if(temp_y == y_north)       north_empty = FALSE;
                else if(temp_y == y_south)  south_empty = FALSE;
            if(temp_y == y)
                if(temp_x == x_east)        east_empty = FALSE;
                else if(temp_x == x_west)   west_empty = FALSE;
        }
```

```
            if(north_empty) {
                site_ptr->xcoord = x;
                site_ptr->ycoord = y_north;
                site_ptr->type = PERIMETER;
                site_ptr++;
                perimeter_length++;
                sites_length++;
            }
            if(south_empty) {
                site_ptr->xcoord = x;
                site_ptr->ycoord = y_south;
                site_ptr->type = PERIMETER;
                site_ptr++;
                perimeter_length++;
                sites_length++;
            }
            if(east_empty) {
                site_ptr->xcoord = x_east;
                site_ptr->ycoord = y;
                site_ptr->type = PERIMETER;
                site_ptr++;
                perimeter_length++;
                sites_length++;
            }
            if(west_empty) {
                site_ptr->xcoord = x_west;
                site_ptr->ycoord = y;
                site_ptr->type = PERIMETER;
                site_ptr++;
                perimeter_length++;
                sites_length++;
            }
        }
        /* Call the MathLink yield function manually. This makes our
           program friendly in the cooperative multitasking environment
           of Macintosh and Windows, and also allows abort requests to
             be processed */
        MLCallYieldFunction(MLYieldFunction(stdlink), stdlink,
                            (MLYieldParameters)0);
    }
    if(MLAbort || sites_length > MAX_SITES - 3) {
        MLPutFunction(stdlink, "Abort", 0);
    } else {
        MLPutFunction(stdlink, "List", cluster_length);
        /* March through sites array, sending back coords of CLUSTER
           sites */
```

```
        for(site_ptr = sites, i = 0; i < cluster_length;
                i++, site_ptr++) {
            while(site_ptr->type != CLUSTER) site_ptr++;
            MLPutFunction(stdlink, "List", 2);
            MLPutInteger(stdlink, site_ptr->xcoord);
            MLPutInteger(stdlink, site_ptr->ycoord);
        }
    }
    free(sites);
}



/*************  seed_random  **************/

void seed_random(int seed) {

    srand((unsigned int) seed);
    MLPutSymbol(stdlink, "Null");
}



/***************  main  ****************/

#if !WINDOWS_MATHLINK
int main(int argc, char *argv[]) {
      return MLMain(argc, argv);
}
#else
int PASCAL WinMain(HANDLE hinstCurrent, HANDLE hinstPrevious,
                   LPSTR lpszCmdLine, int nCmdShow) {

        char  buff[512];
        char FAR * argv[32];
        int argc;

        if(!MLInitializeIcon(hinstCurrent, nCmdShow)) return 1;
        argc = MLStringToArgv(lpszCmdLine, buff, argv, 32);
        return MLMain(argc, argv);
}
#endif
```

E.3 TRAFFIC

```
/****************************************************************


                            onelane.tm


****************************************************************/


/***************** Templates begin *****************/


:Begin:
:Function:        one_lane
:Pattern:         OneLaneC[s_Integer, p_Real, vmax_Integer,
                          maxIterations_Integer]
:Arguments:       {s, p, vmax, maxIterations}
:ArgumentTypes:   {Integer, Real, Integer, Integer}
:ReturnType:      Manual
:End:


:Begin:
:Function:        seed_random
:Pattern:         SeedRandomC[seed_Integer]
:Arguments:       {seed}
:ArgumentTypes:   {Integer}
:ReturnType:      Manual
:End:


/****************** C code begins ******************/


#include "mathlink.h"
#include <stdlib.h>

#define TRUE    1
#define FALSE   0

/* Arbitrary choice outside the range of legal speeds: */
#define EMPTY_SPACE   -1

/* Increment macro that wraps around: */
#define NEXT_INDEX(i) (i != s-1 ? i+1 : 0)

#define MIN(a, b, c) ((tmp = ((a) < (b) ? (a) : (b))) < (c) ? tmp : (c))

void put_road(short *road, int s);
void one_lane(int s, double p, int vmax, int max_iterations);
void seed_random(int seed);
```

```
/**************   one_lane   ***************/

void one_lane(int s, double p, int vmax, int max_iterations) {

    short   i,
            j,
            tmp,
            distance,
            num_iterations,
            speed,
            new_speed,
            *road,
            *next_road,
            *temp_ptr;

    road = (short *) calloc(s, sizeof(short));
    if(!road) {
        MLPutSymbol(stdlink, "$Failed");
        return;
    }
    next_road = (short *) calloc(s, sizeof(short));
    if(!next_road) {
        MLPutSymbol(stdlink, "$Failed");
        free(road);
        return;
    }

    /* Set up initial road configuration */
    temp_ptr = road;
    for(i = 1; i <= s; i++) {
        if( ((double) rand())/((double) RAND_MAX) < p) {
            *temp_ptr++ = (rand()>>8) % (vmax + 1);
        } else {
            *temp_ptr++ = EMPTY_SPACE;
        }
    }

    /* Send back the outer List wrapper, and the first road config */
    MLPutFunction(stdlink, "List", max_iterations + 1);
    put_road(road, s);
```

```
    for(num_iterations =1; num_iterations <= max_iterations && !MLAbort;
            num_iterations++) {
        i = 0;
        while(i < s) {
            speed = road[i];
            if(speed == EMPTY_SPACE) {
                next_road[i] = EMPTY_SPACE;
                            /* won't necessarily stay empty */
                i++;
            } else {
                /* Count how far it is until next car */
                distance = 1;
                j = NEXT_INDEX(i);
                while(road[j] == EMPTY_SPACE) {
                    distance++;
                    j = NEXT_INDEX(j);
                }
                /* Set new speed of this car appropriately */
                new_speed = MIN(vmax, distance - 1, speed + 1);
             /* Move this car ahead that many spaces, marking all the
                    spaces we jump over as empty */
                for(j = i, tmp = 0; tmp < new_speed;
                        j = NEXT_INDEX(j), tmp++) {
                    next_road[j] = EMPTY_SPACE;
                }
                next_road[j] = new_speed;
                i += new_speed + 1;
            }
        }
        put_road(next_road, s);

        temp_ptr = road;
        road = next_road;
        next_road = temp_ptr;
    }

    /* If we finished the loop normally, we've put as many elements as we
       promised, and this MLEndPacket is superfluous. If we finished the
       loop by aborting, then this MLEndPacket is a deliberate error that
        causes $Aborted to be received in Mathematica. */
    MLEndPacket(stdlink);

    free(road);
    free(next_road);
}
```

```
/************* put_road **************/

void put_road(short *road, int s) {

    char  speed;
    int   i;

    MLPutFunction(stdlink, "List", s);
    for(i = 0; i < s; i++) {
        if((speed = road[i]) == EMPTY_SPACE) {
            MLPutSymbol(stdlink, "e");
        } else {
            MLPutInteger(stdlink, speed);
        }
    }
}

/*********** seed_random ***********/

void seed_random(int seed) {

    srand((unsigned int) seed);
    MLPutSymbol(stdlink, "Null");
}

/************* main **************/

#if !WINDOWS_MATHLINK
int main(int argc, char *argv[]) {
    return MLMain(argc, argv);
}
#else
int PASCAL WinMain(HANDLE hinstCurrent, HANDLE hinstPrevious,
                LPSTR lpszCmdLine, int nCmdShow) {

        char  buff[512];
        char FAR * argv[32];
        int argc;

        if(!MLInitializeIcon(hinstCurrent, nCmdShow)) return 1;
        argc = MLStringToArgv(lpszCmdLine, buff, argv, 32);
        return MLMain(argc, argv);
}
#endif
```

E.4 EXCITABLE MEDIA

```
/****************************************************************

                            phases.tm

****************************************************************/


/****************** Templates begin ******************/

:Begin:
:Function:      phases
:Pattern:       PhasesC[s_Integer, n_Integer, maxIterations_Integer]
:Arguments:     {s, n, maxIterations}
:ArgumentTypes: {Integer, Integer, Integer}
:ReturnType:    Manual
:End:

:Begin:
:Function:      seed_random
:Pattern:       SeedRandomC[seed_Integer]
:Arguments:     {seed}
:ArgumentTypes: {Integer}
:ReturnType:    Manual
:End:


/****************** C code begins ******************/

#include "mathlink.h"
#include <stdlib.h>

#define TRUE 1
#define FALSE 0

#define NORTH(i)   ((i) == 1 ? s_sqr - s : -s)
#define EAST(i)    ((i) == s ? 1 - s     : 1 )
#define SOUTH(i)   ((i) == s ? s - s_sqr : s )
#define WEST(i)    ((i) == 1 ? s - 1     : -1)

void phases(int s, int n, int max_iterations);
int put_cell_matrix(char *matrix, int s);
void seed_random(int seed);
```

```
/****************** phases ******************/

void phases(int s, int n, int max_iterations) {

    int    i,
           j,
           index,
           s_sqr,
           num_iterations;
    char   state,
           *cell_matrix,
           *next_cell_matrix,
           *temp_ptr;

    s_sqr = s * s;

    /* Use calloc so that matrix starts out filled with 0's */
    cell_matrix = (char*) calloc(s_sqr, sizeof(char));
    if(!cell_matrix) {
        MLPutSymbol(stdlink, "$Failed");
        return;
    }
    next_cell_matrix = (char*) calloc(s_sqr, sizeof(char));
    if(!next_cell_matrix) {
        MLPutSymbol(stdlink, "$Failed");
        free(cell_matrix);
        return;
    }

    /* Set up the initial random configuration */
    temp_ptr = cell_matrix;
    for(i = 1; i <= s_sqr; i++) {
        *temp_ptr++ = (char) ((rand()>>8) % n);
    }

    MLPutFunction(stdlink, "List", max_iterations+1);
    if(!put_cell_matrix(cell_matrix, s)) MLAbort = TRUE;
```

```
    for(num_iterations = 1; num_iterations <= max_iterations && !MLAbort;
                num_iterations++) {
        index = 0;
        for(i = 1; i <= s; i++) {
            for(j = 1; j <= s; j++, index++) {
                state = cell_matrix[index];
                if(state == n-1) {
                    if(!(cell_matrix[index + NORTH(i)] &&
                          cell_matrix[index + EAST(j)] &&
                          cell_matrix[index + SOUTH(i)] &&
                          cell_matrix[index + WEST(j)])) {
                        next_cell_matrix[index] = 0;
                    } else {
                        next_cell_matrix[index] = state;
                    }
                } else {
                    if(cell_matrix[index + NORTH(i)] == state+1 ||
                        cell_matrix[index + EAST(j)]  == state+1 ||
                        cell_matrix[index + SOUTH(i)] == state+1 ||
                        cell_matrix[index + WEST(j)]  == state+1) {
                        next_cell_matrix[index] = state + 1;
                    } else {
                        next_cell_matrix[index] = state;
                    }
                }
            }
        }
        if(!put_cell_matrix(next_cell_matrix, s)) MLAbort = TRUE;

        temp_ptr = cell_matrix;
        cell_matrix = next_cell_matrix;
        next_cell_matrix = temp_ptr;
    }

    MLEndPacket(stdlink);

    free(cell_matrix);
    free(next_cell_matrix);
}
```

```
/************* put_cell_matrix **************/

int put_cell_matrix(char *cell_matrix, int s) {

    size_t i,
           size,
           matrix_len = s * s;
    char   *str, *p;

    for(i = 0, size = 0; i < matrix_len; i++)
        size += MLPutCharToString(cell_matrix[i], NULL);
    p = str = (char *) malloc(size + 1);
    if(p) {
        for(i = 0; i < matrix_len; i++)
            MLPutCharToString(cell_matrix[i], &p);
        *p = '\0';
        MLPutFunction(stdlink, "Partition", 2);
        MLPutFunction(stdlink, "ToCharacterCode", 1);
        MLPutString(stdlink, str);
        MLPutInteger(stdlink, s);
        free(str);
        return(TRUE);
    } else {
        return(FALSE);
    }
}

/************* seed_random ************/

void seed_random(int seed) {

    srand((unsigned int) seed);
    MLPutSymbol(stdlink, "Null");
}

/*************** main ****************/

#if !WINDOWS_MATHLINK
int main(int argc, char *argv[]) {
    return MLMain(argc, argv);
}
#else
int PASCAL WinMain(HANDLE hinstCurrent, HANDLE hinstPrevious,
                   LPSTR lpszCmdLine, int nCmdShow) {

        char  buff[512];
        char FAR * argv[32];
```

```
        int argc;

        if(!MLInitializeIcon(hinstCurrent, nCmdShow)) return 1;
        argc = MLStringToArgv(lpszCmdLine, buff, argv, 32);
        return MLMain(argc, argv);
}
#endif
```

E.5   SANDPILES

```
/****************************************************************

                         sandpile.tm

****************************************************************/

/*****************  Templates begin  ******************/

:Begin:
:Function:      sandpile
:Pattern:       SandpileC[s_Integer, m_Integer]
:Arguments:     {s, m}
:ArgumentTypes: {Integer, Integer}
:ReturnType:    Manual
:End:

:Begin:
:Function:      seed_random
:Pattern:       SeedRandomC[seed_Integer]
:Arguments:     {seed}
:ArgumentTypes: {Integer}
:ReturnType:    Manual
:End:

/******************  C code begins  ******************/

#include "mathlink.h"
#include <stdlib.h>

#define TRUE   1
#define FALSE  0

void sandpile(int s, int max_iterations);
int put_landscape(char *landscape, int dim);
short count_topheavy_neighbors(int i, int j);
void seed_random(int seed);
```

```
char    *landscape;   /* convenient to have these globals */
size_t  dim;

/**************  sandpile  **************/

void sandpile(int s, int max_iterations) {

    int    i,
           j,
           randx,
           randy,
           done,
           num_iterations;
    char   height,
           *next_landscape,
           *ptr1,
           *ptr2,
           *temp;

    dim = s + 2;
    landscape = (char *) calloc(dim * dim, sizeof(char));
    if(!landscape) {
        MLPutSymbol(stdlink, "$Failed");
        return;
    }
    next_landscape = (char *) calloc(dim * dim, sizeof(char));
    if(!next_landscape) {
        MLPutSymbol(stdlink, "$Failed");
        free(landscape);
        return;
    }

    /* Note pointer increment tricks used here to walk through array,
       skipping over border of zeros that surround the true board */
    ptr1 = landscape + dim + 1;
    for(i = 1; i <= s; i++, ptr1 += 2) {
        for(j = 1; j <= s; j++, ptr1++) {
            *ptr1 = (char) ((rand()>>8) % 2 + 3);
        }
    }

    /* Set up the initial values */
    do {
        randx = (rand()>>4) % s + 1;
        randy = (rand()>>4) % s + 1;
    } while(++landscape[randx * dim + randy] < 5);
```

```
    /* Promise to send the maximum requested number of generations,
       even though the simulation will likely end before this number
       is reached */
    MLPutFunction(stdlink, "List", max_iterations+1);
    if(!put_landscape(landscape, dim)) MLAbort = TRUE;

    num_iterations = 0;
    do {
        done = TRUE;
        num_iterations++;
        ptr1 = landscape + dim + 1;
        ptr2 = next_landscape + dim + 1;
        for(i = 1; i <= s; i++, ptr1 += 2, ptr2 += 2) {
            for(j = 1; j <= s; j++, ptr1++, ptr2++) {
                height = *ptr1;
                if(height <= 4) {
                    *ptr2 = height + count_topheavy_neighbors(i, j);
                } else {
                    done = FALSE;
                    *ptr2 = height + count_topheavy_neighbors(i, j) - 4;
                }
            }
        }

        if(!put_landscape(next_landscape, dim)) MLAbort = TRUE;

        temp = landscape;
        landscape = next_landscape;
        next_landscape = temp;

    } while(!done && num_iterations < max_iterations && !MLAbort);

    if(!MLAbort) {
        /* In case the simulation ended before we got to max_iterations,
           fill out the promised length of the result list by sending
          Sequence[] objects, which will dissolve in place when the list
           hits the evaluator */
        for(i = num_iterations + 1; i <= max_iterations; i++)
           MLPutFunction(stdlink, "Sequence", 0);
    }
    MLEndPacket(stdlink);

    free(landscape);
    free(next_landscape);
}
```

```
/********* count_topheavy_neighbors **********/

#define TOPHEAVYQ(x) ((x) < 5 ? 0 : 1)

/* mult could be eliminated by passing in not i, j, but ptr1-landscape */
short count_topheavy_neighbors(int i, int j) {
    size_t index = i * dim + j;
    return TOPHEAVYQ(landscape[index - dim]) +
           TOPHEAVYQ(landscape[index + 1]) +
           TOPHEAVYQ(landscape[index - 1]) +
           TOPHEAVYQ(landscape[index + dim]);
}


/************** put_landscape **************/

int put_landscape(char *landscape, int dim) {

    size_t i,
           size,
           landscape_len = dim * dim;
    char   *str, *p;

    /* This method is discussed in the section on the PhasesC program */
    for(i = 0, size = 0; i < landscape_len; i++)
        size += MLPutCharToString(landscape[i], NULL);
    p = str = (char *) malloc(size + 1);
    if(p) {
        for(i = 0; i < landscape_len; i++)
            MLPutCharToString(landscape[i], &p);
        *p = '\0';
        MLPutFunction(stdlink, "Partition", 2);
        MLPutFunction(stdlink, "ToCharacterCode", 1);
        MLPutString(stdlink, str);
        MLPutInteger(stdlink, dim);
        free(str);
        return(TRUE);
    } else {
        return(FALSE);
    }
}


/***************** seed_random *****************/
void seed_random(int seed) {

    srand((unsigned int) seed);
    MLPutSymbol(stdlink, "Null");
}
```

```
/*************** main ****************/

#if !WINDOWS_MATHLINK
int main(int argc, char *argv[]) {
    return MLMain(argc, argv);
}
#else
int PASCAL WinMain(HANDLE hinstCurrent, HANDLE hinstPrevious,
                   LPSTR lpszCmdLine, int nCmdShow) {

    char   buff[512];
    char   FAR * argv[32];
    int    argc;

    if(!MLInitializeIcon(hinstCurrent, nCmdShow)) return 1;
    argc = MLStringToArgv(lpszCmdLine, buff, argv, 32);
    return MLMain(argc, argv);
}
#endif
```

E.6   THE GAME OF LIFE

```
/****************************************************************

                          lifegame.tm

****************************************************************/


/****************** Templates begin ******************/

:Evaluate:      LifeGameC::input =
                        "Initial input configuration out of range.
                            Defaulting to random configuration."

:Begin:
:Function:      life
:Pattern:       LifeGameC[s_Integer, maxGens_Integer, initial_List:{}]
:Arguments:     {s, maxGens, If[Max[initial] <= s && Min[initial] >= 1,
                              Flatten[initial],
                              Message[LifeGameC::input]; {}] }
:ArgumentTypes: {Integer, Integer, IntegerList}
:ReturnType:    Manual
:End:
```

```
:Begin:
:Function:       seed_random
:Pattern:        SeedRandomC[seed_Integer]
:Arguments:      {seed}
:ArgumentTypes:  {Integer}
:ReturnType:     Manual
:End:
```

```
/******************  C code begins  ******************/

#include "mathlink.h"
#include <stdlib.h>

#define TRUE    1
#define FALSE   0

#define  ALIVE  1
#define  DEAD   0

#define NORTH(i)           ((i) == 1 ? s_sqr - s : -s)
#define EAST(j)            ((j) == s ? 1 - s      : 1 )
#define SOUTH(i)           ((i) == s ? s - s_sqr : s )
#define WEST(j)            ((j) == 1 ? s - 1      : -1)
#define NORTHEAST(i, j)    NORTH(i) + EAST(j)
#define SOUTHEAST(i, j)    SOUTH(i) + EAST(j)
#define SOUTHWEST(i, j)    SOUTH(i) + WEST(j)
#define NORTHWEST(i, j)    NORTH(i) + WEST(j)

void life(int s, int max_gens, int initial[], long len);
int put_board(char *board, int s);
void seed_random(int seed);


/*********************  life  ********************/

void life(int s, int max_gens, int initial[], long len) {

    long    i,
            j,
            s_sqr;
    int     num_gens,
            done,
            index,
            x_index,
            y_index,
            num_living_neighbors;
```

```
char    state,
        *board,
        *next_board,
        *temp_ptr;


s_sqr = s * s;
/* calloc is used to fill the arrays with 0's initially,
   corresponding to all DEAD sites */
board = (char*) calloc(s_sqr, sizeof(char));
if(!board) {
    MLPutSymbol(stdlink, "$Failed");
    return;
}
next_board = (char*) calloc(s_sqr, sizeof(char));
if(!next_board) {
    MLPutSymbol(stdlink, "$Failed");
    free(board);
    return;
}


/* len is the length of the initial configuration supplied by the
   user. If it is 0, none was supplied, so we default to a random
   one. Otherwise, we set the board up according to the
   specification */
if(len == 0) {
    temp_ptr = board;
    for(i = 0; i < s_sqr; i++, temp_ptr++)
        if((rand()>>8) % 2)
            *temp_ptr = ALIVE;
} else {
    for(i = 0; i < len/2; i++) {
        x_index = *initial++;
        y_index = *initial++;
        board[(y_index - 1) * s + (x_index - 1)] = ALIVE;
    }
}


MLPutFunction(stdlink, "List", max_gens + 1);
if(!put_board(board, s)) MLAbort = TRUE;
num_gens = 0;
```

```
do {
    done = TRUE;
    num_gens++;
    index = 0;
    for(i = 1; i <= s; i++) {
        for(j = 1; j <= s; j++, index++) {
            state = board[index];
            num_living_neighbors =
                      (board[index + NORTH(i)] == ALIVE) +
                    (board[index + NORTHEAST(i, j)] == ALIVE) +
                      (board[index + EAST(j)] == ALIVE) +
                    (board[index + SOUTHEAST(i, j)] == ALIVE) +
                      (board[index + SOUTH(i)] == ALIVE) +
                    (board[index + SOUTHWEST(i, j)] == ALIVE) +
                      (board[index + WEST(j)] == ALIVE) +
                    (board[index + NORTHWEST(i, j)] == ALIVE);
            if(state == ALIVE) {
                if(num_living_neighbors == 2 ||
                   num_living_neighbors == 3) {
                    next_board[index] = ALIVE;
                } else {
                    next_board[index] = DEAD;
                    done = FALSE;
                }
            } else {    /* current state is DEAD */
                if(num_living_neighbors == 3) {
                    next_board[index] = ALIVE;
                    done = FALSE;
                } else {
                    next_board[index] = DEAD;
                }
            }
        }
    }

    if(MLAbort) {
        done = TRUE;
    } else {
        if(!put_board(next_board, s)) {
            MLAbort = TRUE;
            done = TRUE;
        }
    }
    temp_ptr = board;
    board = next_board;
    next_board = temp_ptr;
} while(!done && num_gens < max_gens);
```

```
    if(!MLAbort) {
        /* In case the simulation ended before we got to max_gens,
           fill out the promised length of the result list by sending
           Sequence[] objects, which will dissolve in place when the
           list hits the evaluator */
        for(i = num_gens + 1; i <= max_gens; i++)
            MLPutFunction(stdlink, "Sequence", 0);
    }

    MLEndPacket(stdlink);

    free(board);
    free(next_board);
}


/*************** put_board ***************/

int put_board(char *board, int s) {

    size_t i,
           size,
           board_len = s * s;
    char   *str, *p;

    for(i = 0, size = 0; i < board_len; i++)
        size += MLPutCharToString(board[i], NULL);
    p = str = (char *) malloc(size + 1);
    if(p) {
        for(i = 0; i < board_len; i++)
            MLPutCharToString(board[i], &p);
        *p = '\0';
        MLPutFunction(stdlink, "Partition", 2);
        MLPutFunction(stdlink, "ToCharacterCode", 1);
        MLPutString(stdlink, str);
        MLPutInteger(stdlink, s);
        free(str);
        return(TRUE);
    } else {
        return(FALSE);
    }
}
```

```
/************** seed_random **************/

void seed_random(int seed) {

    srand((unsigned int) seed);
    MLPutSymbol(stdlink, "Null");
}

/**************** main ****************/

#if !WINDOWS_MATHLINK
int main(int argc, char *argv[]) {
    return MLMain(argc, argv);
}
#else
int PASCAL WinMain(HANDLE hinstCurrent, HANDLE hinstPrevious,
                   LPSTR lpszCmdLine, int nCmdShow) {

        char  buff[512];
        char FAR * argv[32];
        int argc;

        if(!MLInitializeIcon(hinstCurrent, nCmdShow)) return 1;
        argc = MLStringToArgv(lpszCmdLine, buff, argv, 32);
        return MLMain(argc, argv);
}
#endif
```

Index

$Aborted, 235
$Failed, 234, 235
:ArgumentTypes:, 224
:Arguments:, 223
:Evaluate:, 251
:Function:, 223
:Pattern:, 223
:ReturnType:, 225

Absorbing boundary conditions, 93
Accretion, 31
addtwo, 223
Algorithmic physics, vii
Animations
    AnimateCA, 168
    AnimateDLA, 42
    AnimateWalk2D, 13
    creation of, 12
    running in notebook, 168
Apply, 174, 198
AspectRatio, 10
Atomic expressions, 174

Ballistic deposition, 31
    algorithm, 38
    fractal dimension, 44
    model, 38
    program (**MolecularDeposition**),
        41
    visualizing (**ShowDeposition**), 41
Beehive life-form (**beehive**), 102

Belousov-Zhabotinsky reaction, 122
Blank (_), 177
BlankNullSequence (___), 177
BlankSequence (__), 177
Borland C++, 227
Boundaries, 93, 94

Canonical ensemble formation, 75
Catastrophes, 105
cc compiler, 228
Cellular automata, 91
    and self-organized criticality, 105
    attractors, 169
    basins of attraction, 169
    catastrophes, 105
    excitable media, 117
    forest fire model, 147
    Game of Life, 97
    Greenberg-Hastings, 117
    Ising Q2R model, 111
    lattices, 91
    liquid to vapor model, 103
    neighborhoods, 92
    number of rules, 160
    rules, 95
    smoothing jagged edges, 103
    spiral CA, 131
    spread of molecules, 103
    traffic, 135
    Wolfram rules, 157
χ^2 (chi-square) test, 211

ClusterLabel, 68

Co-evolution, 83

Compound expressions
  (**CompoundExpression**), 175

Computing remotely, *see* Remote computing

Conway, John, 97

Critical exponent
  of random walk, 8

Crystal growth
  using DLA model, 44

Cyclic space
  algorithm, 120
  C program (**PhasesC**), 243, 276
  debris, 128
  defects, 128
  demons, 129
  droplet enlargement, 131
  droplets, 128
  model, 120
  persistence of phases, 131
  program (**Phases**), 121
  rules, 120
  visualizing, 128

Darwinian evolution
  algorithm, 84
  co-evolution, 83
  distribution of mutations, 88
  model, 83
  program (**DarwinianEvolution**),
    86
  punctuated equilibrium, 83
  theory of, 83

de Gennes, P. G., 57

Diffusion-limited aggregation (DLA), 31
  algorithm, 31
  animating (**AnimateDLA**), 42
  density of, 42
  fractal dimension, 37–38, 42
  model, 31
  modeling crystal growth, 44
  program (**DLA**), 34
  three-dimensional, 43
  visualizing (**ShowDLA**), 34

Distribution of forests, 149–151

DLA, 34

Downvalue, 192

Eden model, 45
  single percolation cluster, 45

Epidemics, *see* Spreading phenomena

Ergodic process, 21

Evaluation, 182

Evolution, *see* Darwinian evolution

Excitable media, 117
  cyclic space, 120
  Hodgepodge machines, 122
  neuron excitation, 117
  visualizing (**ShowExcitation**), 125

Expressions, 173
  atomic, 174
  compound, 175
  entering, 175
  nonatomic, 174

External functions
  calling with *MathLink*, 221

Fibonacci numbers, 190

FixedPoint, 203

FixedPointList, 203

Flow transition, 135

Fold, 201

FoldList, 201

Forest fire
  algorithm, 147
  distribution, 149–151
  model, 147
  program (**SmokeyTheBear**), 149
  self-organized critical state, 152

Forests
  average size, 152

Fractal dimension
  ballistic deposition, 44
  computation of (**FractalDimension**),
    38
  of DLA, 37–38, 42
  of self-avoiding walk, 28

FullForm, 173, 175

Function, 196

Functions, 196
  anonymous, 196
  built-in, 183
  higher-order, 198
  installable, 222
  nested, 203

nested anonymous, 204

Game of Life, 97
    algorithm, 97
    beehive life-form, 102
    C program (**LifeGameC**), 250, 284
    glider life-form, 102
    life-forms, 102
    model, 97
    program (**LifeGame**), 100
    rules, 97
    visualizing (**ShowLife**), 101
Glider life-form (**glider**), 102
GNU C compiler (**gcc**), 228
Graphics
    arrays (**GraphicsArray**), 12
    options, 10
Greenberg-Hastings CA, 117

Head, 175
Hodgepodge machine
    algorithm, 123
    model, 122
    program (**Hodgepodge**), 124
    rules, 122
    visualizing, 130
HoldAll, 188
Hoshen-Kopelman algorithm, 59, 60

Install, 222, 224, 226, 228
Installable functions, 222
IntegerList, 251
Invasion percolation
    algorithm, 49
    model, 49
    program (**Invasion**), 52
Ising model, 75
    algorithm, 75
    animating, 81
    canonical ensemble formation, 75
    clusters, 81
    magnetization of, 80
    probabilistic, 75
    probabilistic vs. deterministic, 116
    program (**IsingMetropolis**), 79
    spin, 75
    spin exchange dynamics, 81

    visualizing (**ShowIsingMagnetization**), 80
Ising Q2R
    model, 111
    program (**IsingCA**), 115
    rules, 111

Kinematic waves, in traffic, 139
Kinetic growth, *see* Spreading phenomena

Lattice
    neighborhood, 92
    rectangular, 91
Lattice walk, 5
Life game, *see* Game of Life
Linear congruential method, 208
LinkObject, 224
LinkPatterns, 225, 227–229
Listable, 200
LogLogListPlot, 9

Malignant arrhythmia, 117
malloc, 235
Manual, 225
Map, 199
MapThread, 200
MatchQ, 178
math, 256
Mathematica, vii
    advantages of, 229
    and diverse programming styles, 230
    high-level programming, 230
    interpreted environment, 229
    memory management, 230
    programming language, xii
    starting from a shell (**math**), 256
    typeless variables, 230
Mathematica programming, 173
    anonymous functions, 196
    atomic expressions, 174
    compound expressions, 175
    constraints on rewrite rules, 189
    constraints on transformation rules, 195
    dynamic rewrite rules, 190
    evaluation, 182
    expressions, 173

functions, 196
higher-order functions, 198
localizing rewrite rules, 189
nested anonymous functions, 204
nested functions, 203
nonatomic expressions, 174
ordering rewrite rules, 191
pattern-matching, 178
pattern-matching sequences, 180
patterns, 177, 181
rewrite rules, 183
rewrite rules and symbols, 192
transformation rules, 193
user-defined rewrite rules, 183
MathLink, 221
aborting external functions, 235
building executables, 224
compiling, 226
developing simulations with, 229
efficiency considerations, 232
efficiency issues, 229
implementing stopping conditions, 247
installing programs, 226, 228
introduction to, 222
launching external programs, 224
lists of arbitrary size, 234
Macintosh platform, 226
memory allocation, 235
optimizing, 229, 243
returning functions, 226
returning results, 225
sample programs, 223
seeding initial configurations, 250
sending multidimensional arrays, 233
specifying arguments, 231
template files, 222, 223
Unix platform, 228
Windows platform, 227
MathLink Developer's Kit, 226
MathLink programs
EpidemicC, 234, 267
LifeGameC, 250, 284
OneLaneC, 240, 272
PhasesC, 243, 276
SandpileC, 247, 280
SeedRandomC, 234
Walk2DC, 231, 263

Matrices
rotation, 22
mcc, 228
command line options, 224
Mean of random walk, 5
MeanSquareDistance, 6
MeanSquareRadiusGyration, 7
Melt, 103
MLAbort, 235
MLCallYieldFunction, 236
MLMain, 223
MLPutCharToString, 244
MLPutFunction, 226, 231
MLPutInteger, 225
MLPutShortIntegerArray, 233
Modeling, vii
Module, 189
MolecularDeposition, 41
Moore neighborhood, 92
mprep, 222
MPW, 226

Nanostructures
formation of, 74
Nearest neighbor, 92
Neighborhood, 92
absorbing boundaries, 93
Moore, 92
periodic boundaries, 93
skewed boundaries, 94
von Neumann, 92
Nest, 202
NestList, 202
Neuron excitation
algorithm, 118
model, 117
program (**NeuronExcitation**), 119
rules, 118
visualizing, 125
nohup, 261
NULL, 233

Off-lattice walk, 14

Part, 174
Partition, 12
Patterns, 177

alternative matching, 181

matching, 178

matching sequences, 180

Percolation, 57

cluster labeling (**ClusterLabel**), 68

cluster-related quantities, 59

Hoshen-Kopelman algorithm, 59, 60

mean cluster size, 72

random site, 57

visualizing, 68

Percolation cluster

algorithm, 46

model, 45

program (**Epidemic**), 49

Periodic boundary conditions, 93

implementing in C, 243

Phases, 121

Physics

algorithmic, vii

Pivot algorithm

for self-avoiding walk, 21

Pivot2DSAW, 27

PlotBinData, 216

PlotRange, 13

Power law, 8

Probability distributions, 213

arbitrary, 218

log normal distribution, 217

normal distribution, 214

Programming

functional, xii, 196

with C, 221

Pseudorandom numbers, *see* Random

numbers

Punctuated equilibrium, 83

Q2R model, 111

Radius of gyration

of random walk, 7

rand(), 230

Random, 209–210

Random number generators

C program (**SeedRandomC**), 234

linear congruential method, 208

quality of, 230

seed for, 207, 212

tests for, 210

Random numbers, 207

and *Mathematica*, 209

χ^2 (chi-square) test, 211

complex, 210

computer generated, 207

from probability distributions, 213

integers, 209

real numbers, 209

sequences of, 207

Random site percolation, 57

Random walk

AnimateWalk2D, 13

animations, 12

C program (**Walk2DC**), 231, 263

critical exponent, 8

mean occupancy, 14

mean value, 5

MeanSquareDistance, 6

numerical analysis, 5

off-lattice, 14

on a lattice, 5

on cubic lattice, 14

one-dimensional (**Walk1D**), 3

radius of gyration, 7

self-avoiding (SAW), 17

sites visited, 14

size, 6

SquareDistance, 6

three-dimensional, 14

two-dimensional, 5

two-dimensional (**Walk2D**), 5

visualizing, 10

$RandomState, 212

RealList, 251

Remote computing, xii, 255

across networks, 259

connections, 257

security considerations, 257

set-up, 256

transferring files, 259, 260

uninterrupted (**nohup**), 261

with local front end, 256

ReplacePart, 174

ReplaceRepeated (**//.**), 195

Rewrite rules, 183

associating with symbols, 192

constraints, 189
dynamic creation of, 190
localizing, 189
ordering, 191
user-defined, 183
Rotation matrices, 22
Rucker, Rudy, 103
Rug, 103
Rule (->), 193
RuleDelayed (:>), 194
Russell, Bertrand, 207

Sandpile
algorithm, 106
C program (**SandpileC**), 247, 280
model, 105
program (**Sandpile**), 107
rules, 105
visualizing (**ShowCatastrophe**), 108
SeedRandom, 212
Self-avoiding walk (SAW), 17
cul-de-sac, 20
efficiency issues, 21
fractal dimension, 28
on cubic lattice, 29
pivot algorithm, 21
Pivot2DSAW, 27
possible pitfalls, 20
shape of, 29
slithering snake, 18
Self-organized criticality, 105, 117
forest fires, 152
Sequence, 247
Set, 183, 184
SetDelayed, 183, 187
Shock waves
in traffic, 139
ShowCA, 165
ShowCatastrophe, 108
ShowDeposition, 41
ShowDLA, 34
ShowDLA2, 35
ShowDLA3D, 44
ShowExcitation, 125
ShowIsingMagnetization, 80
ShowLife, 101
ShowPercolation, 70

ShowSpread, 52
ShowTraffic, 138
ShowWalk2D, 10
Skewed boundary conditions, 94
Slithering snake algorithm, 18
SlitheringSAW, 20
SmokeyTheBear, 149
Spin, 75
Spin exchange dymamics, 81
Spiral cellular automata, 131
SpiralCA, 133
Spreading phenomena, 45
animating, 54
C program (**EpidemicC**), 234, 267
Eden model, 45
kinetic growth, 45
Epidemic, 49
visualizing (**ShowSpread**), 52
with obstacles, 54
SquareDistance, 6
StepIncrements, 3

Template files
with *MathLink*, 222
Think C, 226
Traffic, 135
C program (**OneLaneC**), 240, 272
flow vs. density, 143
fundamental diagram, 143
kinematic waves, 139
lane changes, 144
mean velocity, 143
one-lane algorithm, 135
one-lane model, 135
one-lane program (**OneLane**), 138
shock waves, 139
two-lane algorithm, 140
two-lane model, 140
two-lane program (**TwoLane**), 142
visualizing (**ShowTraffic**), 138
with obstacles, 144
Transferring files, 259, 260
Transformation rules, 193
constraints on, 195
repeated application of, 195

Unix, 228

Upvalue, 192

von Neumann neighborhood, 92
VoteNearCallsToLosers, 104

WaitNextEvent, 236
Walk, *see* Random walk
Wolfram cellular automata, 157
  algorithm, 162
  animating (**AnimateCA**), 168
  program (**WolframCA**), 164
  rules, 157
  visualizing (**ShowCA**), 165
WolframCA, 164